An Introduction to Queueing Theory

An Introduction to Queueing Theory

Brian D. Bunday *BSc, PhD, CMath, FIMA, CStat*

Head of the Department of Mathematics, University of Bradford, UK

A member of the Hodder Headline Group
LONDON • SYDNEY • AUKLAND
Copublished in the Americas by Halsted Press an imprint of John Wiley & Sons Inc.
New York – Toronto

First published in Great Britain in 1996 by
Arnold, a member of the Hodder Headline Group,
338 Euston Road, London, NW1 3BH

Copublished in the Americas by Halsted Press,
an imprint of John Wiley & Sons, Inc.
605 Third Avenue,
New York NY 10158–0012

British Library Cataloguing in Publication Data
A catalogue record for this book is available from the British Library

Library of Congress Cataloging-in-Publication Data
A catalog record for this book is available from the Library of Congress

ISBN 0 340 66239 5
ISBN 0 470 23613 2 (in the Americas only)

Typeset in 10/12 Times by
Mathematical Composition Setters Ltd, Salisbury, UK
Printed and bound in Great Britain by
J. W. Arrowsmith Ltd, Bristol

Contents

Preface

The majority of this book consists of the set of 'hand-outs' for a course in Queueing Theory given to final-year students as part of the syllabus in Operational Research. The course comprised 24 1-hour lectures (two each week) along with 12 1-hour examples classes (one each week). The sets of examples are also included, along with their solutions, which were given to the students after a suitable interval, during which time they attempted the exercises with some hints and help.

The students had a reasonable background in statistics and probability as well as some knowledge of the calculus and matrix algebra. The course was not seen as one in which mathematical rigour was in any way sacrosanct. Indeed the important feature was the intuitive idea of mathematical modelling to illustrate how a variety of situations can be brought successfully within the scope of elementary methods.

Students on the course had the advantage of hearing the many asides and explanations which accompanied and hopefully added to the written material. Readers will not have this advantage but it is hoped that the pace and approach will none the less allow them to assimilate the ideas and techniques. A few additional sections have been added to the original hand-outs in order to assist this process. As far as the examples are concerned readers should make a real effort to solve these problems before having recourse to the solutions provided. The examples show some of the situations in which queueing theory can be useful. The solutions to the examples contain, in addition, a number of insights into the working of queueing systems. In general, material for one lecture precedes each set of examples except for Examples 5, 13, 14, 15. Here two lectures are needed to cope with all the material tested in the examples.

Simulation had already featured in previous courses taken by these students. However, to help the reader and also to give a more complete and overall coverage of the elementary methods used in the analysis of queueing systems, Chapter 10 has been added. This gives a brief résumé of simulation models and their application to queueing problems.

I am grateful to the many students who over the years have had this material presented to them. I hope that others, perhaps working on their own, may also be able to share in the course, and I trust they will find the topics as fascinating as they appear to me.

Brian Bunday
1996

Introduction

All of us have had experience of queueing systems. Commonplace examples include:

- patients at the dentist waiting for treatment;
- customers at a supermarket check-out;
- aeroplanes circling an airport waiting to land;
- telephone calls to an exchange waiting for the operator.

No doubt readers can readily think of other situations.

There are three basic elements in a queueing system.

1. The arrivals: to describe these we need to consider the statistical patterns of the arrivals. Possibilities include regular arrivals or irregular arrivals with the inter-arrival times having some specified distribution. In such cases we can define the rate of arrivals or the average number of arrivals per minute.
2. The service mechanism: to describe this we need to know (a) the number of servers, and (b) the duration of a service time (a random variable in general with a specified distribution).
3. The service discipline: generally first in first out (FIFO), but others are possible in which certain customers get priority. An obvious example would be accident victims who are very seriously injured. They will jump the queue to get treatment.

We study queues in order to be able to predict the effect of changes on the system before we actually implement those changes. The latter may be a very costly exercise. We can perhaps change:

- the pattern of arrivals;
- the mean length of service;
- the number of servers.

We might ask how such changes will affect

- the time customers have to wait (on average);
- the number of customers waiting (on average);
- the proportion of time the service facility is in use.

We now look at the first two elements in isolation, but in more detail.

The arrivals

A key concept is the idea of a Poisson stream or completely random arrivals at a mean rate α per unit time. During the interval $(t, t + \delta t)$ of length δt the probability of an arrival is $\alpha \delta t + o(\delta t)$ and the probability of no arrival is $1 - \alpha \delta t + o(\delta t)$. Here $o(\delta t)$ is a quantity which $\to 0$ faster than δt so that

$$\underset{\delta t \to 0}{\text{Limit}} \; \frac{o(\delta t)}{\delta t} = 0$$

The number of arrivals during an interval of length x is then a random variable denoted by N. Then N has a Poisson distribution with mean αx.

$$\Pr(N = n) = \frac{e^{-\alpha x}(\alpha x)^n}{n!}, \qquad n = 0, 1, 2, \ldots$$

$$N = 3 \text{ in this realisation.}$$

Proof
Divide the interval x into m equal parts each of length x/m $(=\delta t)$ and then let $m \to \infty$ (i.e. $\delta t \to 0$). A customer arrives in n of these parts, no customer arrives in the other $m - n$ parts.

$$\Pr(N = n) = p_n = \underset{m \to \infty}{\text{Limit}} \binom{m}{n}\left(\frac{\alpha x}{m}\right)^n \left(1 - \frac{\alpha x}{m}\right)^{m-n}$$

$$= \underset{m \to \infty}{\text{Limit}} \; \frac{m(m-1)\ldots(m-n+1)}{m^n} \frac{(\alpha x)^n}{n!} \left(1 - \frac{\alpha x}{m}\right)^m \left(1 - \frac{\alpha x}{m}\right)^{-n}$$

$$= \frac{e^{-\alpha x}(\alpha x)^n}{n!} \tag{0.1}$$

This is essentially the proof that the Poisson distribution can arise as the limiting form of a binomial distribution.

$$E[N] = \alpha x, \qquad E[N^2] = \alpha x + (\alpha x)^2; \qquad \text{Var}[N] = \alpha x$$

The probability of more than one arrival during δt is $1 - p_0 - p_1$ where p_0, p_1 are calculated for the interval δt.

$$\therefore \; \Pr\{\text{more than one arrival in } (t, t + \delta t)\} = 1 - p_0 - p_1$$

$$= 1 - e^{-\alpha \delta t} - \frac{(\alpha \delta t)}{1} e^{-\alpha \delta t}$$

$$= 1 - \left(1 - \alpha \delta t + \frac{\alpha^2 \delta t^2}{2!} - \cdots\right) - \alpha \delta t\left(1 - \alpha \delta t + \frac{\alpha^2 \delta t^2}{2!} - \cdots\right)$$

$$= \frac{(\alpha \delta t)^2}{2} + \cdots = o(\delta t) \tag{0.2}$$

The time between two *successive* arrivals is a random variable T.
T has probability density function $f(t) = \alpha e^{-\alpha t}; \; t \geqslant 0$.

Proof

$$\Pr(t \leqslant T \leqslant t + \delta t) = f(t)\delta t \text{ (by definition of the density function)}$$
$$= \Pr(\text{no arrivals in } (0, t)) \times \Pr(\text{one arrival in } \delta t)$$
$$= e^{-at} \times a\delta t$$
$$\therefore f(t) = ae^{-at}; \qquad t \geqslant 0 \tag{0.3}$$

$$E[T] = \frac{1}{a}; \qquad \text{Var}[T] = \frac{1}{a^2}$$

Note:

1. a is the arrival *rate*; a per minute. $1/a$ is the mean interval between successive arrivals, the mean inter-arrival time.
2. The argument above also applies to the interval from an arbitrary moment in time until the instant at which the next arrival occurs. This also has a negative exponential distribution with mean $1/a$. (See question 6 of Examples 1.)

The service mechanism

The duration of service is in general a random variable.

-   If the service time is a constant c this is a random variable with mean c, variance zero and probability density function $f(t) = \delta(t - c)$ where $\delta(\cdot)$ is the Dirac delta function.
-   In many of our models we shall assume that the service time has a negative exponential distribution with parameter β.

Thus if T is the random variable denoting length of service, T has probability density function

$$f(t) = \beta e^{-\beta t}; \qquad t \geqslant 0$$

$$E[T] = \frac{1}{\beta}; \qquad \text{Var}[T] = \frac{1}{\beta^2}$$

In analogy with the arrivals, β is the service rate (while service is going on).

$$\Pr[T \geqslant t] = \int_t^\infty \beta e^{-\beta x} \, dx = e^{-\beta t} \tag{0.4}$$

The probability that a customer's service finishes during an interval of length δx, i.e. in $(x, x + \delta x)$ given that service was in progress at time x, is

$$\Pr(x \leqslant T \leqslant x + \delta x \,|\, T \geqslant x) = \frac{\Pr(x \leqslant T \leqslant x + \delta x)}{\Pr(T \geqslant x)}$$

$$= \frac{\beta e^{-\beta x} \, \delta x}{e^{-\beta x}} = \beta \delta x \text{ (N.B. independent of } x) \tag{0.5}$$

Then in line with the derivation of (0.1), the number of services completed in an interval of length z is a random variable denoted by N, which has a Poisson distribution with mean βz. If the server is busy during the whole interval

$$\Pr(N = n) = \frac{e^{-\beta z}(\beta z)^n}{n!} \tag{0.6}$$

It is important to realise that results (0.4)–(0.6) are only valid for negative exponential service times with mean $1/\beta$.

The meaning of the mathematical results

A key element of this course concerns mathematical modelling. In the context of queueing theory this concerns the way in which the physical working of a queueing system is transformed into a set of mathematical relationships or equations. On solving these equations we obtain further mathematical relationships and by the reverse process these can be interpreted in terms of the physical properties of the system.

We have examined the idea of random arrivals. The mathematical analysis shows that for the three statements:

1. the probability of one arrival in $(t, t + \delta t)$ is $\alpha\delta t + o(\delta t)$ and the probability of no arrivals in $(t, t + \delta t)$ is $1 - \alpha\delta t + o(\delta t)$;
2. the number of arrivals during an interval of length x is a random variable N which has a Poisson distribution with parameter αx;
3. the inter-arrival time between successive customers has a negative exponential distribution with mean $1/\alpha$,

any one implies the other two. Starting with (1) above we have shown (2) at (0.1) and (3) at (0.3).

In order to illustrate the idea of random arrivals we consider the case where patients arrive at random to see their doctor at an average rate of one every 10 minutes. There is no appointments system. Unfortunately the doctor has been called out on an emergency and is 2 hours late in getting to the surgery. Thus for 120 minutes we have a stream of Poisson arrivals to the surgery who on arrival wait for the doctor. In this situation $\alpha = 1/10$ (of a customer per minute). Thus $1/\alpha = 10$ (minutes) is the mean inter-arrival time. In order to illustrate what happens we use statement (3) to get the computer to generate the inter-arrival times of the successive patients. These have been rounded to the nearest minute and are recorded below.

$$11, 1, 2, 5, 2, 1, 8, 9, 6, 4, 18, 4, 7, 9, 28, 1, 9$$

We have in effect got the computer to simulate the arrivals. Using the inter-arrival times we can generate the actual arrival times of the successive patients as 11, $11 + 1 = 12$, $12 + 2 = 14$, 19, 21, 22, 30, 39, 45, 49, 67, 71, 78, 87, 115, 116, 125. Thus when the doctor arrives after 120 minutes, 16 patients are waiting. We suppose that subsequent patients are told of the delay and are asked to come the next day. We can illustrate these arrival times as points on a line.

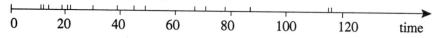

0	20	40	60	80	100	120 time

The times of arrival can be shown to be 16 values chosen at random from 1 to 120, although that is not proved here. Intuitively, statement (1) implies that the probability of an arrival in the next instant is the same at all times. It does not matter that there has just been an arrival or that the last arrival was 10 minutes earlier, say. This is also echoed in note (2) following equation (0.3). The time we have to 'wait' until the next patient arrives always has a negative exponential distribution with mean 10 minutes ($1/\alpha$ in general) irrespective of the previous history. This is the so-called Markov property and completely random arrivals or Poisson arrivals are sometimes also referred to as Markovian arrivals.

Thus, although we get a 'steady' unending stream of arrivals, they occur at random, if that is not a contradiction in terms. In the present realisation we have 16 patients arriving in the first 120 minutes. Had the patients arrived at regular intervals of 10 minutes, say, we should have had 12 arrivals occurring at times 10, 20, 30, ..., 110, 120. This does not present any uncertainty in the analysis. In the present realisation there is irregularity in the pattern of arrivals with a relatively large number arriving early and fewer at the later stages. We shall see that it is this random irregularity which leads to congestion in a queueing system. [See question 1 of Examples 5.]

Of course this latter point is covered by statement (2). If we divide the 120 minutes into three intervals each of 40 minutes then statement (2) says that the number of arrivals in each 40-minute interval will have a Poisson distribution with mean 4. In our realisation we have eight patients in the first, five in the second and three in the third interval.

For the service mechanism we can carry out a similar simulation. Suppose the time the doctor spends with a patient has a negative exponential distribution with mean 5 minutes $(1/\beta$ in our notation). We can get the computer to generate the 16 service times for our 16 patients. Rounded up these are:

$$2, 3, 10, 3, 2, 1, 5, 4, 11, 5, 3, 10, 7, 15, 5, 1$$

In this case we can separate and analyse separately the arrival and service patterns. The doctor will start to see his patients at time 120 minutes and the 'service completion' times for the patients will thus be:

$$122, 125, 135, 138, 140, 141, 146, 150, 161, 166, 169, 179, 186, 201, 206 \text{ and } 207$$

The service times are 'inter-departure times' since service is going on continuously until all patients have been seen. Again we can represent these departure times as points on a line. Had the service time been constant of duration 5 minutes, the 16 departure times would have been at 125, 130, 135, ..., 195, 200.

Again we see the departure points randomly spread on the line in contrast to what it would have been with constant service.

Analogous to random arrivals, for negative exponential service times with mean $1/\beta$, when service is ongoing we have for the three statements

(a) the service times of successive customers are independent and have an exponential distribution with mean $1/\beta$;
(b) the probability that a service ongoing at time x will finish in $(x, x + \delta x)$ is $\beta \delta x + o(\delta x)$, and will not finish in $(x, x + \delta x)$ is $1 - \beta \delta x + o(\delta x)$;
(c) the number of service completions in time z when service is continuous has a Poisson distribution with mean βz.

Any one statement implies the other two.

Starting from (a) we have shown (b) at (0.5) and (c) at (0.6). In connection with (b) it is important to realise that it does not matter for how long the service has been in progress. The probability that the service ends in the next instant δx is $\beta \delta x$ to first order in δx. Analogous to note (2) for arrivals, it is also true that for service which is in progress the residual service time, i.e. the remaining time until the completion of the service, will have a negative exponential distribution with mean $1/\beta$, again independent of the length of time for which the service has been in progress. [See question 7 of

Examples 1.] Again it is emphasised that this is only true for negative exponential service times.

With regard to (c) in the intervals (120, 160), (160, 200) we have 8 and 5 service completions and these are realisations of a Poisson variable with mean $8(1/5 \times 40)$. This rather unusual, artificial situation has allowed us to examine the arrival process and the service process in isolation. When the two proceed side by side we have a genuine queueing system and we shall examine this in more detail later.

Finally we make a brief comment on (0.2). This says that the probability of two 'simultaneous' arrivals in δt is of the order of magnitude $(\delta t)^2$. Thus it is 'doubly infinitesimally' small. In continuous time we will not have simultaneous events. They will always be separated by at least δt if we make δt small enough.

Examples 1

1 Customers arrive at a location and congregate there. Customers arrive at random at an average rate α. If at time zero there are already i customers at the location and $p_n(t)$ is the probability that there are n customers at the location at time t, show that

$$p_n(t) = \frac{e^{-at}(at)^{n-i}}{(n-i)!} \quad \text{for } n \geq i$$

2 For random arrivals at rate α, X_k denotes the length of time until the kth arrival. Show that X_k is a random variable with probability density function (p.d.f.)

$$f(x) = \frac{a(ax)^{k-1}e^{-ax}}{(k-1)!}$$

[Note if $k = 1$ we have the time to the first arrival; (0.3).]

3 The service time X has p.d.f. $\varphi(x) = \beta e^{-\beta x}$.
Show that $\varphi(x)$ has Laplace transform

$$E[e^{-sX}] = \varphi^*(s) = \frac{\beta}{\beta + s}$$

Hence deduce the mean and variance of X.

4 Use the result concerning the Laplace transform in question 3 to give an alternative solution to question 2.

5 If the service time X is a constant b and so has p.d.f. $f(x) = \delta(x - b)$, then:
(i)   sketch the density function of X;
(ii) sketch the distribution function of X.

6 Consider a stream of random arrivals at average rate α. Consider an arbitrary moment in time and the wait T from that time until the next arrival. Show that T has a negative exponential distribution with mean $1/\alpha$.

7 Consider a moment during the service time of a customer whose service time has a negative exponential distribution with mean $1/\beta$. The service of this customer is ongoing. Consider the residual service time X of this customer, i.e. X is the remaining time until the service is completed. Show that X also has a negative exponential distribution with mean $1/\beta$.

Solutions to Examples 1

1 We need an additional $(n - i)$ arrivals

$$p_n(t) = \Pr((n - i) \text{ arrivals in time } t) = \frac{e^{-at}(at)^{n-i}}{(n-i)!}$$

2
$$f(x)\delta x = \Pr(x \leqslant X_k \leqslant x + \delta x) \qquad \text{by definition of the p.d.f.}$$
$$= \Pr(\text{exactly } k - 1 \text{ arrivals by time } x)$$
$$\times \Pr(\text{one arrival in } (x, x + \delta x))$$
$$= \frac{(ax)^{k-1}e^{-ax}}{(k-1)!} \times a\delta x$$

whence, since this is true for arbitrary small δx,

$$f(x) = \frac{a(ax)^{k-1}e^{-ax}}{(k-1)!}$$

3
$$\varphi^*(s) = \int_0^\infty e^{-sx}\beta e^{-\beta x}\, dx = \beta \int_0^\infty e^{-(s+\beta)x}\, dx = \frac{\beta}{s+\beta}$$

$$\varphi^*(s) = \left(1 + \frac{s}{\beta}\right)^{-1} = \left(1 - \frac{s}{\beta} + \frac{s^2}{\beta^2} - \frac{s^3}{\beta^3} + \cdots\right)$$

$$E[X] = -\text{coeff. } s = \frac{1}{\beta}$$

$$E[X^2] = \text{coeff. } \frac{s^2}{2!} = \frac{2}{\beta^2}$$

$$\therefore \; \text{Var}[X] = E[X^2] - \{E[X]\}^2 = \frac{2}{\beta^2} - \frac{1}{\beta^2} = \frac{1}{\beta^2}$$

$$\text{Note: } E[X] = -\frac{d\varphi^*(s)}{ds}\bigg|_{s=0} \qquad E[X^2] = (-1)^2 \frac{d^2\varphi^*(s)}{ds^2}\bigg|_{s=0}$$

4 If T_i denotes the interval between the $(i-1)$th and the ith arrival then
$$X_k = T_1 + T_2 + \cdots + T_k$$
where the T_i are independent random variables which are identically distributed with p.d.f.
$$h(t) = ae^{-at} \text{ and Laplace transform } h^*(s) = \frac{a}{a+s}$$

Thus if X_k has p.d.f. $f(x)$ and this has Laplace transform $f^*(s)$ then

$$f^*(s) = [h^*(s)]^k = \frac{a^k}{(a+s)^k}$$

This is readily seen to be the Laplace transform of

$$f(x) = \frac{a(ax)^{k-1}e^{-ax}}{(k-1)!}$$

For

$$\int_0^\infty \frac{a(ax)^{k-1}e^{-ax}e^{-sx}}{(k-1)!}dx = \int_0^\infty \frac{a^k x^{k-1}e^{-(a+s)x}}{(k-1)!}dx$$

$$= \frac{a^k}{(a+s)^k}\int_0^\infty \frac{u^{k-1}e^{-u}}{(k-1)!}du$$

on making the substitution $u = (a+s)x$

$$= \left(\frac{a}{a+s}\right)^k$$

5

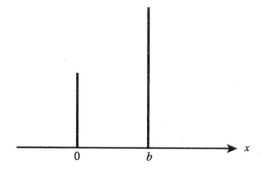

Density function

$f(x)$ is zero everywhere except when $x = b$ where there is a 'spike' of area 1.

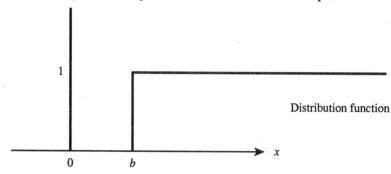

Distribution function

The distribution function is a Heaviside unit-step function.

6 We follow the argument which leads to (0.3)

$\Pr(t \leqslant T \leqslant t + \delta t) = f(t)\delta t$ by definition of the density function $f(t)$

$= \Pr(\text{no arrivals in } (0,\, t)) \times \Pr(\text{one arrival in } \delta t)$

$= e^{-\alpha t} \times \alpha \delta t$

where we have used (0.1) with $n = 0$.
Since this is true for arbitrarily small δt, $f(t) = \alpha e^{-\alpha t}$.

7 We measure X from the arbitrary moment when service is ongoing and suppose that X has probability density function $h(x)$. Then

$\Pr(x \leqslant X \leqslant x + \delta x) = h(x)\delta x$

$= \Pr(\text{no service completions in } (0,\, x))$
$\quad \times \Pr(\text{service completion in } \delta x)$

$= e^{-\beta x} \times \beta \delta x$

where we have used (0.6) with $n = 0$.

$\therefore h(x) = \beta e^{-\beta x}$

so that X has a negative exponential distribution with mean $1/\beta$.

1
Generating functions

1.1 The probability generating function

We shall find the idea of the probability generating function a useful tool in the analysis of queueing systems. Probability generating functions are widely used in the study of stochastic processes and queueing systems are special examples of such processes. We digress briefly to mention some of the important results and applications.

If N is a discrete random variable which can assume the values n $(= 0, 1, 2, \ldots)$ with probability p_n then we define the probability generating function as

$$P(z) = \mathrm{E}[z^N] = \sum_{n=0}^{\infty} p_n z^n \tag{1.1}$$

$P(z)$ is a convergent series and a well-behaved function of z for $|z| \leqslant 1$. Of course

$$P(1) = \sum_{n=0}^{\infty} p_n = 1 \tag{1.2}$$

Sometimes the 'dummy variable' is denoted by s and we have

$$P(s) = \sum_{n=0}^{\infty} p_n s^n$$

The z can be complex and sometimes it is necessary to bear this in mind. Sometimes the way to find p_n is to find $P(z)$ and expand the latter as a power series in z. Then p_n is the coefficient of z^n. Note that $p_0 = P(0)$. If $'$ denotes differentiation with respect to z then

$$P'(z) = \sum_{n=0}^{\infty} n p_n z^{n-1}$$

so that

$$P'(1) = \mathrm{E}[N] = \sum_{n=0}^{\infty} n p_n \tag{1.3}$$

Similarly

$$P''(1) = \mathrm{E}[N(N-1)] = \sum_{n=0}^{\infty} n(n-1) p_n \tag{1.4}$$

Example 1.1

If N is the score on a die when it is thrown at random then

$$\Pr(N = n) = p_n = \tfrac{1}{6} \qquad \text{for } n = 1, 2, 3, 4, 5, 6$$

Thus from (1.1) in this case

$$P(z) = \tfrac{1}{6}z + \tfrac{1}{6}z^2 + \tfrac{1}{6}z^3 + \tfrac{1}{6}z^4 + \tfrac{1}{6}z^5 + \tfrac{1}{6}z^6$$

It is clear that

$$P(1) = \tfrac{1}{6} + \tfrac{1}{6} + \cdots + \tfrac{1}{6} = 1$$

Also

$$P'(z) = \tfrac{1}{6}[1 + 2z + 3z^2 + 4z^3 + 5z^4 + 6z^5]$$

and it is clear that $P'(1)$ is term by term $\Sigma n\, p_n$ as at (1.3), and so gives the expected score as $21/6 = 7/2$. Of course in this case we can write

$$P(z) = \frac{z}{6}\left[\frac{z^6 - 1}{z - 1}\right]$$

on summing the geometric series.

If we try to evaluate $P(1)$ in this form we get an indeterminate form $0/0$. We have to resolve this using l'Hopital's rule.

$$\left[\text{If } \underset{z \to a}{\text{Limit}}\ \frac{R(z)}{S(z)} \text{ takes the form } \frac{0}{0}, \text{ then } \underset{z \to a}{\text{Limit}}\ \frac{R(z)}{S(z)} = \underset{z \to a}{\text{Limit}}\ \frac{R'(z)}{S'(z)}\right]$$

In this example

$$P(1) = \underset{z \to 1}{\text{Limit}}\ \frac{1}{6}\left(\frac{z^6 - 1}{z - 1}\right)\left(= \frac{0}{0}\right)$$

$$= \underset{z \to 1}{\text{Limit}}\ \frac{1}{6}\frac{6z^5}{1}$$

$$= 1$$

Similar problems will arise when we try to calculate the mean as $P'(1)$. (See question 2 of Examples 2.)

As an alternative $P'(1) = E[N]$ is the coefficient of $y\ [=(z-1)]$ in the expansion of $P(z)$ as a power series in y, i.e. $(z - 1)$. For if

$$P(z) = \sum_{n=0}^{\infty} p_n z^n = \sum_{n=0}^{\infty} a_n(z - 1)^n = \sum_{n=0}^{\infty} a_n y^n$$

$$\frac{dP}{dz} = \frac{dP}{dy} \cdot \frac{dy}{dz} = \frac{dP}{dy}$$

since $dy/dz = 1$ if $y = z - 1$.

$$\therefore \quad \frac{dP}{dz} = \sum_{n=0}^{\infty} np_n z^n = \sum_{n=0}^{\infty} na_n(z-1)^{n-1} = \sum_{n=0}^{\infty} na_n y^{n-1}$$

But if $z = 1$, then $y = 0$.

$$\therefore \quad \frac{dP}{dz}\bigg|_{z=1} = \Sigma np_n = a_1$$

since all other terms in $\sum_{n=0}^{\infty} na_n(z-1)^{n-1}$ are zero when $z = 1$ except the term when n is 1.

Example 1.2

N has a binomial distribution with parameters m and p.

$$p_n = \Pr(N = n) = \binom{m}{n} p^n (1-p)^{m-n}, \qquad n = 0, 1, 2, \ldots, m$$

$$\therefore \quad P(z) = \sum_{n=0}^{m} \binom{m}{n} p^n z^n (1-p)^{m-n} = [pz + (1-p)]^m$$

$$E[N] = P'(z)\big|_{z=1} = mp[pz + 1 - p]^{m-1}\big|_{z=1}$$

$$= mp$$

There is no difficulty here. The other method also works very well for

$$P(z) = [pz + 1 - p]^m = [1 + p(z-1)]^m = [1 + py]^m$$

$$= 1 + m(py) + \frac{m(m-1)}{2}(py)^2 + \cdots + (py)^m$$

and the coefficient of y is just mp.

1.2   The moment generating function – the Laplace transform

The Laplace transform of the probability function (discrete case) or the probability density function (continuous case) is defined for the random variable X (non-negative) to be

$$f^*(s) = E[e^{-sX}] = \sum_{n=0}^{\infty} p_n e^{-ns} \qquad \text{discrete case } (X \equiv N)$$

$$= \int_0^{\infty} f(x) e^{-sx} \, dx \qquad \text{continuous case} \qquad (1.5)$$

$$f^*(s) = \int_0^{\infty} e^{-sx} f(x) \, dx = \int_0^{\infty} \left(1 - sx + \frac{s^2 x^2}{2!} - \frac{s^3 x^3}{3!} + \cdots\right) f(x) \, dx$$

$$= 1 - \mu_1 s + \mu_2 \frac{s^2}{2!} - \mu_3 \frac{s^3}{3!} + \cdots$$

where

$$\mu_r = \int_0^\infty x^r f(x) \, dx = E[X^r]$$

$$= (-1)^r \left. \frac{d^r f^*(s)}{ds^r} \right|_{s=0} \tag{1.6}$$

Of course

$$\mu_1 = E[X] = -\text{coeff. of } s$$

$$\mu_2 = E[X^2] = \text{coeff. of } \frac{s^2}{2}$$

$$\text{Var}[X] = \mu_2 - \mu_1^2$$

Theorem
If X_1 and X_2 (≥ 0) are two *independent* random variables with Laplace transforms $f_1^*(s)$ and $f_2^*(s)$ and $Y = X_1 + X_2$ has p.d.f. $\varphi(y)$ and Laplace transform $\varphi^*(s)$ then

$$\varphi^*(s) = f_1^*(s) f_2^*(s) \tag{1.7}$$

Proof

$$\varphi^*(s) = E[e^{-sY}] = E[e^{-s(X_1 + X_2)}]$$
$$= E[e^{-sX_1} \cdot e^{-sX_2}] = E[e^{-sX_1}]E[^{-sX_2}]$$

since X_1 and X_2 are independent

$$\therefore \; \varphi^*(s) = f_1^*(s) f_2^*(s)$$

Note $\varphi(y) = \int_0^y f_1(x) f_2(y - x) dx$ so that

$$\varphi^*(s) = \int_0^\infty e^{-sy} \varphi(y) dy = f_1^*(s) f_2^*(s) \qquad \text{by the convolution theorem.}$$

$$\varphi^*(s) = \int_0^\infty \left(\int_0^y f_1(x) f_2(y - x) dx \right) e^{-sy} dy$$

$$= \int_0^\infty \left(\int_x^\infty f_1(x) f_2(y - x) e^{-sy} dy \right) dx$$

on changing the order of integration (*see* Fig. 1.1).
    If we put $v = y - x$ in the 'inner integral'

$$\varphi^*(s) = \int_0^\infty f_1(x) \left(\int_0^\infty f_2(v) e^{-s(v + x)} dv \right) dx = \int_0^\infty e^{-sx} f_1(x) dx \int_0^\infty e^{-sv} f_2(v) dv$$

$$= f_1^*(s) f_2^*(s)$$

This constitutes a second proof of the result.

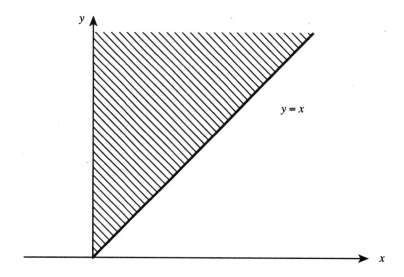

Fig. 1.1

What we have to realise is that the probability distribution, the probability generating function and the Laplace transform all contain the information about the random variable. Sometimes the latter can be a more useful vehicle for conveying and using this information.

Example 1.3

Service to patients at a clinic consists of two independent stages. Stage 1 comprises an examination and stage 2 treatment. The time to carry out each stage has a negative exponential distribution with probability density function

$$f(t) = \beta e^{-\beta t}; \qquad t \geq 0$$

Find the density function for the total service time and verify the correctness of (1.7) in this case.

The total service time, $T = T_1 + T_2$
T_1 and T_2 both have probability density function

$$f(t) = \beta e^{-\beta t}; \qquad t \geq 0$$

Each has Laplace transform

$$f^*(s) = \int_0^\infty \beta e^{-\beta t} e^{-st}\, dt$$

$$= \int_0^\infty \beta e^{-(\beta + s)t}\, dt$$

$$= \frac{\beta}{\beta + s}$$

Thus T has probability density function $\varphi(t)$, say, where as noted

$$\varphi(t) = \int_0^t \beta e^{-\beta x} \beta e^{-\beta(t-x)} \, dx$$

$$= e^{-\beta t} \int_0^t \beta^2 \, dx$$

$$\therefore \quad \varphi(t) = \beta^2 t e^{-\beta t}$$

Thus

$$\varphi^*(s) = \int_0^\infty \varphi(t) e^{-st} \, dt$$

$$= \int_0^\infty \beta^2 t e^{-(\beta+s)t} \, dt$$

In this integral put $z = (\beta + s)t$, so that $t = z/(\beta + s)$ and $dt = dz/(\beta + s)$.

$$\therefore \quad \varphi^*(s) = \frac{\beta^2}{(\beta+s)^2} \int_0^\infty z e^{-z} \, dz = \frac{\beta^2}{(\beta+s)^2} = \left(\frac{\beta}{\beta+s} \right)^2$$

It is clear that $\varphi^*(s) = f^*(s) . f^*(s)$ as required.

1.3   Examples 2

1   The random variable N which has probability function

$$\Pr(N = n) = p_n, \qquad \text{for } n = 0, 1, 2, \ldots$$

has probability generating function

$$P(z) = \frac{k}{6 - z - z^2}$$

Find k and hence determine p_0 and verify that $p_0 = P(0)$. Show that

$$p_n = \frac{4}{10} \left(\frac{1}{2} \right)^n + \frac{4}{15} \left(-\frac{1}{3} \right)^n$$

[Hint: Write $P(z)$ in partial fractions prior to a power series expansion of the separate terms.]
Hence evaluate $E[N]$ as $P'(1)$ or as the sum of a series.

N.B.   If $|x| < 1$ then $\displaystyle\sum_{n=0}^\infty nx^n = \frac{x}{(1-x)^2}$

2   Use the generating function in the form

$$P(z) = \frac{z}{6} \left(\frac{z^6 - 1}{z - 1} \right)$$

to find the expected score when a standard die is thrown. Use both methods suggested.

3 Analogous to the theorem concerning the Laplace transform of the sum of two independent variables is a theorem concerning generating functions. If N_1 and N_2 are two independent discrete random variables taking the values 0, 1, 2, ... and with probability generating functions $P_1(z)$ and $P_2(z)$, respectively, and $M = N_1 + N_2$ has probability generating function $R(z)$, then

$$R(z) = P_1(z)P_2(z)$$

Prove this result.

Suppose two standard dice are thrown simultaneously and let T denote the total score. Find the probability generating function for T, the probability that the total score is 4, and the expected value of the total score.

4 The integer-valued random variable N has probability generating function

$$P(z) = \frac{K(z-1)}{z\, e^{\rho(1-z)} - 1} \qquad \text{where } \rho < 1$$

Show that $K = 1 - \rho$ and hence find $p_0 = \Pr(N = 0)$. Find the value of $E[N]$.

1.4   Solutions to examples 2

1
$$P(z) = \frac{k}{6 - z - z^2} = \frac{k}{(3+z)(2-z)}$$

Since
$$P(1) = 1, \frac{k}{4} = 1 \quad \therefore \quad k = 4$$

$$p_0 = P(0) = \frac{4}{6} = \frac{2}{3}$$

$$P(z) = \frac{4}{(3+z)(2-z)} = \frac{4}{5}\left[\frac{1}{3+z} + \frac{1}{2-z}\right]$$

$$= \frac{4}{10}\left(1 - \frac{z}{2}\right)^{-1} + \frac{4}{15}\left(1 + \frac{z}{3}\right)^{-1}$$

$$= \frac{4}{10}\left(1 + \frac{z}{2} + \frac{z^2}{2^2} + \frac{z^3}{2^3} + \cdots\right) + \frac{4}{15}\left(1 - \frac{z}{3} + \frac{z^2}{3^2} - \frac{z^3}{3^3} + \cdots\right)$$

$$\therefore \quad p_0 = \frac{4}{10} + \frac{4}{15} \quad (\text{coeff. of } z^0)$$

$$= \frac{12}{30} + \frac{8}{30} = \frac{20}{30} = \frac{2}{3}$$

$$p_n = \text{coeff. of } z^n = \frac{4}{10}\left(\frac{1}{2}\right)^n + \frac{4}{15}\left(-\frac{1}{3}\right)^n$$

$$E[N] = \sum_{n=0}^{\infty} np_n = \frac{4}{10} \sum_{n=0}^{\infty} n\left(\frac{1}{2}\right)^n + \frac{4}{15} \sum_{n=0}^{\infty} n\left(-\frac{1}{3}\right)^n$$

$$= \frac{4}{10} \cdot \frac{\frac{1}{2}}{\left(1-\frac{1}{2}\right)^2} + \frac{4}{15} \cdot \frac{\left(-\frac{1}{3}\right)}{\left(1+\frac{1}{3}\right)^2}$$

$$= \frac{4}{10} \times 2 - \frac{4}{15} \cdot \frac{3}{16} = \frac{8}{10} - \frac{3}{60} = \frac{45}{60} = \frac{3}{4}$$

But $P'(z) = \dfrac{4}{5} \left[-\dfrac{1}{(3+z)^2} + \dfrac{1}{(2-z)^2} \right]$

$$\therefore \quad E[N] = P'(1) = \frac{4}{5}\left[1 - \frac{1}{16}\right] = \frac{4}{5} \times \frac{15}{16} = \frac{3}{4}$$

This second method seems to be easier.

2

$$P(z) = \frac{z}{6}\left(\frac{z^6-1}{z-1}\right) = \frac{1}{6}\left(\frac{z^7-z}{z-1}\right)$$

Using the quotient rule for differentiation we obtain

$$P'(z) = \frac{1}{6} \cdot \frac{(z-1)(7z^6-1)-(z^7-z)1}{(z-1)^2} = \frac{1}{6} \cdot \frac{6z^7-7z^6+1}{z^2-2z+1}$$

$\therefore \quad P'(1)$

$$= \operatorname*{Limit}_{z \to 1} \frac{1}{6} \cdot \frac{6z^7-7z^6+1}{z^2-2z+1} \quad \left(=\frac{0}{0}\right)$$

$$= \operatorname*{Limit}_{z \to 1} \frac{1}{6} \cdot \frac{42z^6-42z^5}{2z-2} \quad \left(=\frac{0}{0}\right) \qquad \text{[l'Hopital]}$$

$$= \operatorname*{Limit}_{z \to 1} \frac{1}{6} \cdot \frac{252z^5-210z^4}{2} = \frac{42}{12} = \frac{7}{2} \qquad \text{[l'Hopital again]}$$

Thus the expected score is $7/2$.

Alternatively with $y = z - 1$ (or $z = y + 1$),

$$P(z) \text{ as a function of } y \text{ is } \frac{1}{6}\left[\frac{(y+1)^7-(y+1)}{y}\right]$$

$$P(z) = \frac{1}{6}\left[\frac{6y+21y^2+\cdots+y^7}{y}\right] = \frac{1}{6}[6+21y+\cdots+y^6]$$

and the coefficient of y in this expansion is $21/6 = 7/2$ and this is the expected score. This method seems to have considerable advantage over the double application of l'Hopital's rule.

3
$$M = N_1 + N_2$$
$$R(z) = E[Z^M] = E[Z^{N_1 + N_2}] = E[Z^{N_1} Z^{N_2}]$$
$$= E[Z^{N_1}] E[Z^{N_2}]$$

since N_1 and N_2 are independent.
But

$$E[Z^{N_1}] = P_1(z) \text{ and } E[Z^{N_2}] = P_2(z)$$
$$\therefore \quad R(z) = P_1(z) P_2(z)$$

For our problem $T = S_1 + S_2$ where S_1 and S_2 are the scores on the two dice. They are clearly independent random variables and as we have seen from Example 1 of this section both have probability generating function

$$P_1(z) = P_2(z) = \tfrac{1}{6}(z + z^2 + z^3 + z^4 + z^5 + z^6)$$
$$\therefore \quad R(z) = \tfrac{1}{36}(z + z^2 + z^3 + z^4 + z^5 + z^6)(z + z^2 + z^3 + z^4 + z^5 + z^6)$$
$$= \tfrac{1}{36}(z + z^2 + z^3 + z^4 + z^5 + z^6)^2$$

The probability that $T = 4$ is the coefficient of z^4 in this expansion. The term in z^4 arises from products $z \times z^3$, $z^2 \times z^2$ and $z^3 \times z$ in the two bracketed expressions and corresponds to the pairs $(S_1, S_2) \equiv (1, 3)$, $(2, 2)$ or $(3, 1)$. The coefficient of z^4 is $3/36 = 1/12$ and is $\Pr(T = 4)$.

$$R'(z) = \tfrac{2}{36}(z + z^2 + z^3 + z^4 + z^5 + z^6)(1 + 2z + 3z^2 + 4z^3 + 5z^4 + 6z^5)$$

Thus

$$E[T] = R'(1) = \tfrac{2}{36} . 6 \times 21 = 2 \times \tfrac{7}{2} = 7$$

as we would expect since $E[T] = E[S_1] + E[S_2]$.

4 $P(1) = 1$, but $P(1)$ takes the form $0/0$.
The denominator $z e^{\rho(1-z)} - 1 = 0$ when $z = 1$. Thus $P(z)$ is in fact well behaved when $z = 1$ as indeed it must be.
We have to use l'Hopital's rule

$$\underset{z \to 1}{\text{Limit}} \, P(z) = \underset{z \to 1}{\text{Limit}} \, \frac{K}{e^{\rho(1-z)} - \rho z \, e^{\rho(1-z)}} = \frac{K}{1 - \rho} = 1$$

$$\therefore \quad K = (1 - \rho)$$

$$\therefore \quad P(z) = \frac{(1 - \rho)(z - 1)}{z \, e^{\rho(1-z)} - 1}$$

$$p_0 = P(0) = \frac{-K}{-1} = K = (1 - \rho)$$

One way to find $E[N]$ is to evaluate $P'(z)$ when $z = 1$. This has problems since the denominator is zero when $z = 1$.

$$P'(z) = \frac{(1-\rho)[(ze^{\rho(1-z)} - 1) - (z-1)(e^{\rho(1-z)} - \rho ze^{\rho(1-z)})]}{[ze^{\rho(1-z)} - 1]^2}$$

As $z \to 1$ this takes a $0/0$ form. If we differentiate top and bottom we get

$$\frac{(1-\rho)\{e^{\rho(1-z)} - \rho ze^{\rho(1-z)} - (e^{\rho(1-z)} - \rho ze^{\rho(1-z)}) - (z-1)[\rho^2 ze^{\rho(1-z)} - 2\rho e^{\rho(1-z)}]\}}{2[ze^{\rho(1-z)} - 1][e^{\rho(1-z)} - \rho ze^{\rho(1-z)}]}$$

$$= \frac{-(1-\rho)(z-1)[\rho^2 ze^{\rho(1-z)} - 2\rho e^{\rho(1-z)}]}{2[ze^{\rho(1-z)} - 1][e^{\rho(1-z)} - \rho ze^{\rho(1-z)}]}$$

Now as $z \to 1$ $\dfrac{(1-\rho)(z-1)}{ze^{\rho(1-z)} - 1} \to 1$ since $P(1) = 1$.

Thus

$$E[N] = -\frac{1}{2} \operatorname*{Limit}_{z \to 1} \frac{[\rho^2 ze^{\rho(1-z)} - 2\rho e^{\rho(1-z)}]}{[e^{\rho(1-z)} - \rho ze^{\rho(1-z)}]}$$

$$= -\frac{1}{2} \frac{\rho^2 - 2\rho}{1-\rho} = \frac{2\rho - \rho^2}{2(1-\rho)}$$

This is a case where it might be easier to expand $P(z)$ as a power series in y ($= z - 1$) and find the coefficient of y. As we have seen, this also will give $E[N]$.

As a function of y, $P(z) = \dfrac{(1-\rho)y}{(1+y)e^{-\rho y} - 1} = Q(y)$ say;

$$Q(y) = \frac{(1-\rho)y}{(1+y)\left(1 - \rho y + \dfrac{\rho^2 y^2}{2} - \dfrac{\rho^3 y^3}{3!} + \cdots\right) - 1} = \frac{(1-\rho)y}{(1-\rho)y + y^2\left(\dfrac{\rho^2}{2} - \rho\right) + \cdots}$$

$$= \frac{1}{1 + y\dfrac{(\rho^2 - 2\rho)}{2(1-\rho)} + y^2( )}$$

$$= \left[1 - y\frac{(2\rho - \rho^2)}{2(1-\rho)} + y^2( )\right]^{-1}$$

$$= 1 + y\frac{(2\rho - \rho^2)}{2(1-\rho)} + \cdots \qquad \text{as far as the first power of } y.$$

Thus $E[N] = $ coeff. of y in this expansion

$$= \frac{2\rho - \rho^2}{2(1-\rho)} \quad \text{as before.}$$

1.5   Random arrivals – a generating function approach

Consider a situation where we have a Poisson stream of random arrivals at rate λ. On arrival these customers just 'hang around'. Our model describes what happens in the infinitesimal interval $(t, \ t + \delta t)$. The mathematical solution then allows us to understand what happens in a finite interval, say $(0, \ t)$ in general.

$$\Pr(\text{an arrival in } (t, \ t + \delta t)) = \lambda \delta t + o(\delta t)$$

$$\Pr(\text{no arrival in } (t, \ t + \delta t)) = 1 - \lambda \delta t + o(\delta t)$$

$$\Pr(\text{more than 1 arrival in } (t, \ t + \delta t)) = o(\delta t)$$

Let $p_n(t) = \Pr(\text{there are } n \text{ arrivals by time } t)$. We can suppose we start with 0 arrivals present at time $t = 0$, i.e.

$$p_n(0) = 0 \qquad \text{if } n \neq 0, \qquad p_0(0) = 1$$

$$p_0(t + \delta t) = p_0(t)\,(1 - \lambda \delta t + o(\delta t))$$

There are none there at $t + \delta t$ if there are none at t and no arrival occurs in $(t, \ t + \delta t)$.

$$p_1(t + \delta t) = p_0(t)\,(\lambda \delta t + o(\delta t)) + p_1(t)\,(1 - \lambda \delta t + o(\delta t))$$

There is one there at $t + \delta t$ if there are none at t and one arrives in $(t, \ t + \delta t)$ or there is one at t and no arrival in $(t, \ t + \delta t)$.

Thus,

$$p_n(t + \delta t) = p_{n-1}(t)\,(\lambda \delta t + o(\delta t)) + p_n(t)\,(1 - \lambda \delta t + o(\delta t))$$

$$\therefore \quad \frac{p_0(t + \delta t) - p_0(t)}{\delta t} = -\lambda p_0(t) + \frac{o(\delta t)}{\delta t}$$

$$\frac{p_1(t + \delta t) - p_1(t)}{\delta t} = \lambda p_0(t) - \lambda p_1(t) + \frac{o(\delta t)}{\delta t} \qquad \text{etc.}$$

$$\frac{p_n(t + \delta t) - p_n(t)}{\delta t} = \lambda p_{n-1}(t) - \lambda p_n(t) + \frac{o(\delta t)}{\delta t}$$

[N.B. Since $p_n(t)$ is bounded $p_n(t)o(\delta t)$ is just $o(\delta t)$].

We now let $\delta t \to 0$. The L.H.S. become derivatives and by definition,

$$\underset{\delta t \to 0}{\text{Limit}} \ \frac{o(\delta t)}{\delta t} \to 0$$

Thus,

$$\frac{dp_0(t)}{dt} = -\lambda p_0(t)$$

$$\frac{dp_1(t)}{dt} = \lambda p_0(t) - \lambda p_1(t)$$

$$\frac{dp_2(t)}{dt} = \lambda p_1(t) - \lambda p_2(t) \tag{1.8}$$

$$\frac{dp_n(t)}{dt} = \lambda p_{n-1}(t) - \lambda p_n(t) \qquad \text{etc.}$$

The modelling process wherein we have considered what can happen in a typical interval $(t, t + \delta t)$ has resulted in the equations (1.8). Of course the $p_n(t)$ describe in a probabilistic manner the way the system behaves. To make further progress we need to solve these equations. One way to solve them is to define the generating function

$$\pi(z, t) = \sum_{n=0}^{\infty} p_n(t) z^n$$

This is the probability generating function for the $p_n(t)$. It also involves t since the probabilities change in time. If we multiply the equations (1.8) in turn by z^0, z^1, z^2, etc. and sum we obtain

$$\sum_{n=0}^{\infty} z^n \frac{dp_n(t)}{dt} = \lambda z \sum_{n=0}^{\infty} z^n p_n(t) - \lambda \sum_{n=0}^{\infty} z^n p_n(t)$$

$$\therefore \quad \frac{\partial \pi}{\partial t} = \lambda(z - 1)\pi \tag{1.9}$$

and the *set* of simultaneous equations (1.8) for the $p_n(t)$ have been replaced by *one* equation (1.9) for the generating function.

$$\pi(z, t) = C e^{\lambda(z-1)t}$$

where C is a constant (perhaps a function of z) is the solution to (1.9).

The initial conditions for the $p_n(t)$ are equivalent to the initial condition $\pi(z, 0) = 1$ $\therefore C = 1$.

$$\therefore \quad \pi(z, t) = e^{-\lambda t} e^{\lambda z t}$$

$$= e^{-\lambda t} \left[1 + \lambda t z + \frac{(\lambda t z)^2}{2!} + \frac{(\lambda t z)^3}{3!} + \cdots \right]$$

whence

$$p_n(t) = \text{coeff. of } z^n = \frac{e^{-\lambda t}(\lambda t)^n}{n!}$$

Thus if $\{N(t)\}$ is the random variable (stochastic process) which represents the number of arrivals by time t

$$\Pr\{N(t) = n\} = p_n(t) = \frac{e^{-\lambda t}(\lambda t)^n}{n!}, \qquad \text{Poisson with mean } \lambda t.$$

$E[N(t)] = \lambda t$ as we know or we can obtain this from the generating function

$$\pi(z, t) = e^{-\lambda t} e^{\lambda z t}$$

$$\frac{\partial \pi}{\partial z}(z, t) = \lambda t\, e^{-\lambda t} e^{\lambda z t} = \sum_{n=0}^{\infty} n p_n(t) z^{n-1}$$

and

$$E[N(t)] = \sum_{n=0}^{\infty} n p_n(t) = \left. \frac{\partial \pi}{\partial z} \right|_{z=1} = \lambda t$$

Thus our mathematical analysis has led to the same results and physical interpretation of these derived results as in equations (0.1) and (0.2).

1.6 Non-homogeneous Poisson arrivals

If the arrival rate varies with time so that λ is replaced by $\lambda(t)$, then all the modelling holds up to the equation (1.9) for the generating function which now becomes

$$\frac{\partial \pi}{\partial t} = \lambda(t)(z - 1)\pi$$

Thus treating z as a constant (as far as t is concerned)

$$\int \frac{d\pi}{\pi} = (z - 1) \int \lambda(t)\,dt$$

$$\therefore \quad [\ln \pi(z, t)]'_{t=0} = (z - 1) \int_0^t \lambda(t)\,dt$$

$$\therefore \quad \frac{\pi(z, t)}{\pi(z, 0)} = \exp\left\{(z - 1) \int_0^t \lambda(u)\,du\right\}$$

Thus, if as before $\pi(z, 0) = 1$

$$\pi(z, t) = \exp[-\Lambda(t)]\exp[z\Lambda(t)] \tag{1.10}$$

where $\Lambda(t) = \int_0^t \lambda(u)\,du$ [Note if $\lambda(u)$ is a constant λ, then $\Lambda(t) = \lambda t$]

$$\therefore \quad p_n(t) = e^{-\Lambda(t)} \cdot \frac{(\Lambda(t))^n}{n!} \tag{1.11}$$

and the number to arrive has a Poisson distribution with mean $\Lambda(t)$.

The process described by the probabilities (1.11) and generating function (1.10) is called a non-homogeneous Poisson process. In the present context it describes, in a stochastic sense, the arrival pattern of random arrivals when the (instantaneous) arrival rate varies in time according to the form of the function $\lambda(t)$. It does in fact have many other applications, particularly in the area of reliability theory, but this is not discussed here.

1.7 Random arrivals and negative exponential service – again!

Many of the models developed in this course assume that the arrival pattern is adequately modelled by a stream of random arrivals at average rate α. The validity of the models depends on this assumption. We in turn must question whether the assumption is reasonable in the context of the situation being investigated.

We have seen that this assumption implies that the number of arrivals in time x is a random variable having a Poisson distribution with mean αx. We know also that the Poisson distribution arises as a limiting form of the binomial distribution. Thus in the early applications of queueing theory to the stream of incoming telephone calls to an exchange, one might envisage a very large number n of subscribers, each with a small probability p of making a call within a particular 10-minute interval, say, but such that np is finite. These are just the conditions under which the Poisson distribution approximates the binomial. This is, therefore, the type of context in which we might be fairly confident that our assumption is reasonable and of course the early queueing theory developed to help design telephone exchanges made this assumption.

Clearly the model will apply in many other contexts, arrival of victims to an accident department, certain failures of machines in a factory, but we should not just make the assumption without thinking. It will not apply to every queueing system.

In the early applications to telephone calls, the service time was the duration of a call. Observations show that this is reasonably approximated by a negative exponential distribution. A characteristic of the negative exponential distribution with mean c is that it is rather skew and also has standard deviation c. Thus the variation about the mean as compared to the mean is large, and very large values can occur from time to time. Again this will often be the case in other situations but we should not make the assumption blindly.

We shall see that with these assumptions concerning the arrival and service processes we can develop a number of very interesting and useful models which allow us to predict, for example, average queue lengths and waiting times. In fact the models are quite robust to departures from the assumptions but we should take care not to use the models in situations where they clearly do not apply, or where the assumptions underlying the models cannot be justified. In the context of incoming telephone calls, the rate of arrivals in the mornings when many business calls are made may differ very much from that in the evenings when domestic calls are made. Working days will have very different arrival rates from those at the weekend. In that sense the non-homogeneous Poisson arrivals just discussed may be more appropriate.

1.8 Examples 3

1 Consider the problem of describing random arrivals using the generating function approach, but suppose that at time $t = 0$ there are already i customers present. Show in this case that

$$\pi(z, t) = z^i e^{-\lambda t} e^{\lambda z t}$$

Hence deduce an expression for $p_n(t)$. Compare this with question 1 of Examples 1.

2 Consider the non-homogeneous random arrivals considered at the end of this section. If $\lambda(t) = ae^{-t/b}$ find an expression for $\Lambda(t)$.
Show that the probability that no customers arrive during the interval (u, v) is

$$\exp\{-ab(e^{-(u/b)} - e^{-(v/b)})\}$$

3 Let $X_1, X_2, \ldots, X_n$ be the n values in a random sample from a uniform distribution over the range $(0, t)$ with probability density function

$$f(x) = \frac{1}{t}, \qquad 0 \leqslant x \leqslant t$$

Show that the joint density function

$$f(x_1, x_2, \ldots, x_n) = \frac{1}{t^n}; \qquad 0 \leqslant x_i \leqslant t$$

If Y_1, Y_2, ..., Y_n are the sample values in ascending magnitude $0 \leqslant Y_1 \leqslant Y_2 \leqslant ... \leqslant Y_n \leqslant t$, then Y_1, Y_2, ..., Y_n are called the order statistics of the sample. Show that their joint density function is

$$f(y_1, y_2, ..., y_n) = \frac{n!}{t^n}; \qquad 0 \leqslant y_1 \leqslant y_2 \leqslant \cdots \leqslant y_n \leqslant t$$

4 For a system of random arrivals at average rate λ, let the time of the nth arrival be denoted by Y_n. Suppose it is known that $Y_{n+1} = t$. Show that the conditional joint density for Y_1, Y_2, ..., Y_n is

$$f(y_1, y_2, ..., y_n \,|\, t) = \frac{n!}{t^n}; \qquad 0 \leqslant y_1 \leqslant y_2 \leqslant \cdots \leqslant y_n \leqslant t$$

Thus the successive arrival times have the same distribution as the order statistics of a random sample of n values from the uniform distribution on the range $(0, t)$.

5 Sequel to question 4. Suppose it is known that exactly n arrivals have occurred by time t. Show once again that the conditional joint density for Y_1, Y_2, ..., Y_n is

$$f(y_1, y_2, ..., y_n \,|\, t) = \frac{n!}{t^n}; \qquad 0 \leqslant y_1 \leqslant y_2 \leqslant \cdots \leqslant y_n \leqslant t$$

The results established in questions 4 and 5 may help to explain the term 'random arrivals' for this particular pattern. Given that n customers have arrived by time t, their arrival times, when unordered, are uniformly and independently distributed on $(0, t)$. We made reference to this immediately after the diagram representing the arrival of patients to the surgery in the Introduction.

1.9   Solutions to Examples 3

1 We have $\pi(z, t) = C e^{\lambda(z-1)t}$.
Now $p_i(0) = 1$ and $p_n(0) = 0$ for $n \neq i$.
Thus

$$\pi(z, 0) = \sum_{n=0}^{\infty} p_n(0)z^n = z^i$$

$$\therefore \quad C = z^i$$

so that

$$\pi(z, t) = z^i e^{-\lambda t} e^{\lambda z t}$$

$$\therefore \quad \pi(z, t) = z^i e^{-\lambda t}\left[1 + \lambda t z + \frac{(\lambda t)^2 z^2}{2!} + \frac{(\lambda t)^3 z^3}{3!} + \cdots\right]$$

$$\therefore \quad p_n(t) = \text{coeff. of } z^n \text{ is } \frac{e^{-\lambda t}(\lambda t)^{n-i}}{(n-i)!}$$

2
$$\Lambda(t) = \int_0^t a e^{-(t/b)} \, dt = ab(1 - e^{-(t/b)})$$

If we start observing the system at time u and count the customers who arrive after that time then we can suppose that at time u there are none in the system. Then from equation (1.9)

$$\frac{\partial \pi}{\partial t} = \lambda(t)(z - 1)\pi$$

$$\int_{t=u}^{t=t} \frac{d\pi}{\pi} = (z - 1) \int_u^t \lambda(y) dy$$

$$\therefore \quad [\ln \pi(z, t)]_{t=u}^{t=t} = (z - 1) \int_u^t \lambda(y) dy$$

Note this automatically fixes the arbitrary constant since we integrate from time u to time t.

$$\therefore \quad \ln\left[\frac{\pi(z,t)}{\pi(z,u)}\right] = (z - 1) \int_u^t a e^{-(y/b)} \, dy$$

$$\therefore \quad \ln[\pi(z,t)] = (z - 1)ab(e^{-(u/b)} - e^{-(t/b)})$$

since $\pi(z, u) = 1$ because $p_0(u) = 1$ and $p_n(u) = 0$ for $n \neq 0$.

$$\therefore \quad \pi(z, t) = \exp[(z - 1)ab(e^{-(u/b)} - e^{-(t/b)})]$$
$$= \exp[-ab(e^{-(u/b)} - e^{-(t/b)})]\exp[zab(e^{-(u/b)} - e^{-(t/b)})]$$
$$\therefore \quad \pi(z, v) = \exp[-ab(e^{-(u/b)} - e^{-(v/b)})]\exp[zab(e^{-(u/b)} - e^{-(v/b)})]$$

Therefore the probability of no customers by time v, i.e. in the interval (u, v), is the coefficient of z^0 in the above and so is

$$\exp[-ab(e^{-(u/b)} - e^{-(v/b)})]$$

3 $X_1, X_2, ..., X_n$ are independent since they are a random sample from the distribution. Thus

$$f(x_1, x_2, ..., x_n) = f(x_1)f(x_2)f(x_3)...f(x_n) = \left(\frac{1}{t}\right)^n; \qquad 0 \leqslant x_i \leqslant t$$

There are $n!$ different samples which will generate the same order statistics, namely the $n!$ permutations of the values $Y_1, Y_2, ..., Y_n$. Thus

$$f(y_1, y_2, ..., y_n) = \frac{n!}{t^n}; \qquad 0 \leqslant y_1 \leqslant y_2 \leqslant \cdots \leqslant y_n \leqslant t$$

4 The inter-arrival times $y_1, y_2 - y_1, y_3 - y_2, ..., t - y_n$ are independent and have a negative exponential distribution with mean $1/\lambda$. The random variable Y_{n+1} has probability density function

$$\varphi(y) = \frac{\lambda^{n+1} t^n e^{-\lambda t}}{n!}.$$

from the solution to question 2 of Examples 1.

Thus the joint density of Y_1, Y_2, ..., Y_n, given that $Y_{n+1} = t$, is

$$f(y_1, y_2, ..., y_n | t) = \frac{\lambda e^{-\lambda y_1} \lambda e^{-\lambda(y_2 - y_1)} \lambda e^{-\lambda(y_3 - y_2)} ... \lambda e^{-\lambda(t - y_n)}}{\dfrac{\lambda^{n+1} t^n e^{-\lambda t}}{n!}}$$

$$= \frac{n!}{t^n} \qquad \text{for } 0 \leqslant y_1 \leqslant y_2 \leqslant ... \leqslant y_n \leqslant t$$

5 In this case the conditional joint density of Y_1, Y_2, ..., Y_n, given that the number of arrivals by time t is n, is given by

$$f(y_1, y_2, ..., y_n | t) = \frac{\lambda e^{-\lambda y_1} \lambda e^{-\lambda(y_2 - y_1)} ... \lambda e^{-\lambda(y_n - y_{n-1})} e^{-\lambda(t - y_n)}}{\dfrac{(\lambda t)^n e^{-\lambda t}}{n!}}$$

for $0 \leqslant y_1 \leqslant y_2 \leqslant \cdots \leqslant y_n \leqslant t$.

The denominator is the Poisson probability for n arrivals by time t. The numerator is the product of the densities for the n independent inter-arrival times y_1, $y_2 - y_1$, ..., $y_n - y_{n-1}$ followed by the probability that there is no further arrival in the interval $t - y_n$. Thus again

$$f(y_1, y_2, ..., y_n | t) = \frac{n!}{t^n}; \qquad \text{for } 0 \leqslant y_1 \leqslant y_2 \leqslant ... \leqslant y_n \leqslant t$$

2
Simple queues

2.1 A very simple system

An operator works with a machine which breaks down from time to time. When this occurs the operator repairs it and the machine runs again until the next breakdown, etc. If the stops occur at random at an average rate α during *running time* and the length of a repair time has a negative exponential distribution with p.d.f.

$$f(t) = \beta e^{-\beta t}; \qquad t \geq 0$$

we can find expressions for the probabilities that the machine is running or stopped at time t. These describe the stochastic behaviour of the system.

Let $p_R(t) \equiv$ probability (the machine is running at time t) and $p_S(t) \equiv$ probability (the machine is stopped at time t).

There are just two states. During an interval $(t, t + \delta t)$ the transition probabilities are:

$$
\begin{array}{ll}
R \rightarrow R & 1 - \alpha \delta t \\
R \rightarrow S & \alpha \delta t \\
S \rightarrow R & \beta \delta t \\
S \rightarrow S & 1 - \beta \delta t
\end{array}
$$

to first order in δt. Thus terms of $o(\delta t)$ are omitted. Then if we consider the situation during $(t, t + \delta t)$

$$p_R(t + \delta t) = p_R(t)(1 - \alpha \delta t) + p_S(t)\beta \delta t + o(\delta t)$$

The machine is running at $t + \delta t$ if it is running at t and does not break down in $(t, t + \delta t)$ or is stopped at time t and the repair is completed in $(t, t + \delta t)$.

$$p_S(t + \delta t) = p_R(t)\alpha \delta t + p_S(t)(1 - \beta \delta t) + o(\delta t)$$

The machine is stopped at $t + \delta t$ if it is running at t and breaks down in $(t, t + \delta t)$ or it is stopped at time t and its repair is not completed in $(t, t + \delta t)$.

Thus

$$\frac{dp_R(t)}{dt} = -\alpha p_R(t) + \beta p_S(t)$$

$$\frac{dp_S(t)}{dt} = \alpha p_R(t) - \beta p_S(t)$$

$$(2.1)$$

Once again $p_R(t)$ and $p_S(t)$ describe the probabilistic behaviour of the system. By considering what can physically occur in a typical interval $(t, t + \delta t)$ the modelling

process has produced the equations (2.1) for these quantities. We can solve these equations by elementary methods and since $p_R(t) + p_S(t) = 1$
[note $d/dt \{p_R(t) + p_S(t)\} = 0$]

$$p_R(t) = \frac{\beta}{\alpha + \beta} [1 - e^{-(\alpha + \beta)t}] + p_R(0)e^{-(\alpha + \beta)t}$$

(2.2)

$$p_S(t) = \frac{\alpha}{\alpha + \beta} [1 - e^{-(\alpha + \beta)t}] + p_S(0)e^{-(\alpha + \beta)t}$$

where $p_R(0)$ and $p_S(0)$ give the initial probabilities at time zero of being in the respective states. [See Examples 4, question 1 for details.] This gives the time-dependent solution [transient solution]. In this case it is easy to deduce the steady-state solution.

$$p_R = \underset{t \to \infty}{\text{Limit}}\, p_R(t) = \frac{\beta}{\alpha + \beta}$$

(2.3)

$$p_S = \underset{t \to \infty}{\text{Limit}}\, p_S(t) = \frac{\alpha}{\alpha + \beta}$$

Of course if a steady-state solution exists

$$\underset{t \to \infty}{\text{Limit}}\, \frac{dp_R(t)}{dt} = 0 \qquad \underset{t \to \infty}{\text{Limit}}\, \frac{dp_S(t)}{dt} = 0$$

$\therefore$ To find the steady-state solution we have

$$0 = -\alpha p_R + \beta p_S$$
$$0 = \alpha p_R - \beta p_S$$

and

$$p_R + p_S = 1$$

whence

$$p_R = \frac{\beta}{\alpha + \beta}, \qquad p_S = \frac{\alpha}{\alpha + \beta}$$

This is the easy way to find the steady-state solution without first finding the time-dependent solution. This latter task is often quite hard (if not impossible). Plots of $p_R(t)$ and $p_S(t)$ show what is happening (see Fig. 2.1).

Note if we commence with the system in the steady state so that $p_R(0) = p_R$ and $p_S(0) = p_S$ then from the general solution (2.2),

$$p_R(t) = p_R \qquad \text{and} \qquad p_S(t) = p_S \qquad \text{for all } t$$

Furthermore, we note that p_R and p_S are independent of the initial conditions. Also after a long time p_R is the proportion of time the system is in state R, the proportion of time the machine is running. p_S is the proportion of time the machine is stopped. This is also the proportion of time the operator spends on repairs [see Examples 4, question 2). Thus our modelling has resulted in the mathematical equations (2.1) with (2.2) and (2.3) as solutions.

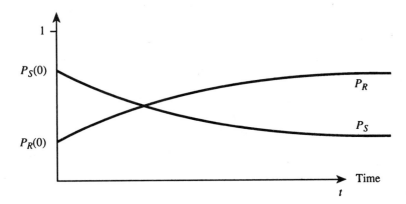

Fig. 2.1

Note the physical interpretation above. However the machine is started, in the long run it reverts to being stopped with probability p_S and runs with probability p_R. As well as being the probabilities of finding the machine in these states (on a random visit at a random time), these probabilities are also the long run proportions of time that the machine is in these states.

Example 2.1

A lead press used in cable manufacture operates 24 hours a day on a shift system. From time to time the machine needs adjustment, the periods between adjustments having a negative exponential distribution with mean 100 minutes. The time taken by the operative to make the adjustment has a negative exponential distribution with mean 10 minutes. Estimate the proportion of time the lead press is operational, and the proportion of time the operative spends carrying out adjustments.

An operative takes over a lead press which is up and running at the start of his 12-hour shift. What is the probability that the machine is running after one hour?

The negative exponential nature of the running time until adjustment means that the machine breaks down at random at an average rate $\alpha = 0.01$. For the repairs, $1/\beta = 10$ so that $\beta = 0.1$ with minutes as the time unit. Thus for the steady-state situation from (2.3)

$$p_R = \frac{0.1}{0.1 + 0.01} \approx 0.909$$

and this represents the proportion of time the machine is running. Also,

$$p_S = \frac{0.01}{0.11} = 0.0909$$

and this represents the proportion of time the operator is actually carrying out adjustments. He is thus left with about 91% of his time free to carry out other tasks.

If the lead press is running at time $t = 0$, then in (2.2) with $p_R(0) = 1$,

$$p_R(t) = \frac{0.1}{0.11} [1 - e^{-0.11t}] + e^{-0.11t}$$

Thus after 1 hour, $t = 60$ and

$$p_R(60) = 0.909 + 0.0909e^{-0.11 \times 60}$$
$$= 0.909 + 0.000123669$$
$$= 0.909 \qquad \text{[to 3 decimal places]}$$

Thus we observe that the steady-state situation is reached for all practical purposes very early on in the shift.

2.2 Examples 4

1 Show that the solution of the equations

$$\frac{dp_R(t)}{dt} = -\alpha p_R(t) + \beta p_S(t)$$

$$\frac{dp_S(t)}{dt} = \alpha p_R(t) - \beta p_S(t)$$

where $p_R(t) + p_S(t) = 1$ is

$$p_R(t) = \frac{\beta}{\alpha + \beta} [1 - e^{-(\alpha + \beta)t}] + p_R(0)e^{-(\alpha + \beta)t}$$

$$p_S(t) = \frac{\alpha}{\alpha + \beta} [1 - e^{-(\alpha + \beta)t}] + p_S(0)e^{-(\alpha + \beta)t}$$

where $p_R(0)$ and $p_S(0)$ are the probabilities that the system is in state R and state S, respectively, when $t = 0$.

2 For the system underlying question 1 show that the average length of a period in state R (the average run time before breakdown) is $1/\alpha$ and the average length of a period in state S (the average duration of a repair) is $1/\beta$.

Consider $2N$ transitions (where N is large). Show that the proportion of time spent in R is $\beta/(\alpha + \beta)$ and in S is $\alpha/(\alpha + \beta)$. These, of course, are p_R and p_S.

3 Telephone calls to an operator arrive at random at an average rate λ. If the operator is busy dealing with an earlier call when a call arrives, that call is lost. The time to deal with a call has a negative exponential distribution with p.d.f. $\mu e^{-\mu t}$; $t \geqslant 0$.

Let $p(t)$ be the probability that the operator is busy dealing with a call at time t. The operator is busy at time 0. Produce an argument to show that

$$p(t) = e^{-\mu t} + \lambda \int_0^t [1 - p(x)]e^{-\mu(t - x)} \, dx$$

(i) From this obtain a differential equation for $p(t)$ and hence find $p(t)$

(ii) Consider the Laplace transform $p^*(s)$. Find $p^*(s)$ and hence $p(t)$.

2.3   Solutions to examples 4

1 $x = p_R(t)$, $y = p_S(t)$

$$\therefore \quad \frac{dx}{dt} = -\alpha x + \beta y \qquad \frac{dy}{dt} = \alpha x - \beta y, \quad x + y = 1$$

$$\therefore \quad \frac{dx}{dt} = -\alpha x + \beta(1 - x) = \beta - (\alpha + \beta)x$$

$$\frac{dx}{dt} + (\alpha + \beta)x = \beta$$

The solution is of the form particular integral + complementary function

$$\therefore \quad x = \frac{\beta}{\alpha + \beta} + Ae^{-(\alpha + \beta)t} \text{ and similarly } y = \frac{\alpha}{\alpha + \beta} + De^{-(\alpha + \beta)t}$$

$$\therefore \quad x(0) = \frac{\beta}{\alpha + \beta} + A \qquad \therefore \quad A = x(0) - \frac{\beta}{\alpha + \beta} \quad [x(0) = p_R(0) \text{ of course}]$$

$$\therefore \quad x = \frac{\beta}{\alpha + \beta} [1 - e^{-(\alpha + \beta)t}] + p_R(0)e^{-(\alpha + \beta)t} = p_R(t)$$

$$y = \frac{\alpha}{\alpha + \beta} [1 - e^{-(\alpha + \beta)t}] + p_S(0)e^{-(\alpha + \beta)t} = p_S(t)$$

2 Periods in R have an exponential distribution with mean $1/\alpha$ and periods in S have an exponential distribution with mean $1/\beta$ (see Introduction and Fig. 2.2). Thus the time spent in R is $N(1/\alpha)$ and the time spent in S is $N(1/\beta)$ asymptotically.

Therefore, the proportion of time spent in R is

$$\frac{\dfrac{N}{\alpha}}{\dfrac{N}{\alpha} + \dfrac{N}{\beta}} = \frac{\dfrac{1}{\alpha}}{\dfrac{1}{\alpha} + \dfrac{1}{\beta}} = \frac{\beta}{\alpha + \beta}$$

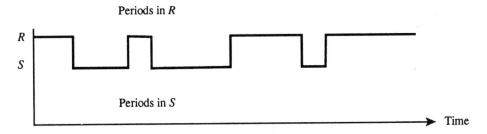

Fig. 2.2

and the proportion of time spent in S is

$$\frac{\dfrac{N}{\beta}}{\dfrac{N}{\alpha} + \dfrac{N}{\beta}} = \frac{\alpha}{\alpha + \beta}$$

3 The operator is busy at time 0. Since the time to deal with a call has a negative exponential distribution the probability that the time to deal with this exceeds t is $e^{-\mu t}$. Thus, $p(t) = \text{Pr(time to deal with first call exceeds } t)$ (OR and hence +)
+ Pr(operator is free at time x *and* a call arrives in $(x, x + dx)$ *and* time to deal with this exceeds $(t - x)$ for all values of x between 0 and t).

$$\therefore\ p(t) = e^{-\mu t} + \int_0^t (1 - p(x))\lambda e^{-\mu(t-x)}\,dx$$

$$p(t) = e^{-\mu t} + \lambda e^{-\mu t} \int_0^t (1 - p(x))e^{\mu x}\,dx$$

[The modelling derives an equation for $p(t)$. Can we solve it?]

$$\therefore\ \frac{dp(t)}{dt} = -\mu e^{-\mu t} - \lambda\mu e^{-\mu t} \int_0^t (1 - p(x))e^{\mu x}\,dx + \lambda e^{-\mu t}(1 - p(t))e^{\mu t}$$

$$= -\mu e^{-\mu t} - \mu[p(t) - e^{-\mu t}] + \lambda(1 - p(t))$$

$$\therefore\ \frac{dp(t)}{dt} = \lambda - (\lambda + \mu)p(t)$$

$$\therefore\ p(t) = \frac{\lambda}{\lambda + \mu} + Ae^{-(\lambda + \mu)t}$$

But $p(0) = 1$ and therefore $A = \mu/(\lambda + \mu)$.

$$\therefore\ p(t) = \frac{\lambda}{\lambda + \mu} + \frac{\mu e^{-(\lambda + \mu)t}}{\lambda + \mu}$$

As an alternative using Laplace transforms

$$p(t) = e^{-\mu t} + \lambda \int_0^t (1 - p(x))e^{-\mu(t-x)}\,dx$$

We note that the integral on the right is the convolution integral for $(1 - p(x))$ and $e^{-\mu x}$. Thus if $p^*(s)$ is the Laplace transform of $p(t)$ since the Laplace transform of 1 is $1/s$ and of $e^{-\mu t}$ is $1/(\mu + s)$ we have

$$p^*(s) = \frac{1}{\mu + s} + \lambda\left[\frac{1}{s} - p^*(s)\right]\frac{1}{\mu + s}$$

$$\therefore\ p^*(s)\left[1 + \frac{\lambda}{\mu + s}\right] = \frac{1}{\mu + s} + \frac{\lambda}{s(\mu + s)}$$

$$\therefore p^*(s)[\mu + s + \lambda] = 1 + \frac{\lambda}{s}$$

$$\therefore p^*(s) = \frac{1}{\mu + \lambda + s} + \frac{\lambda}{s(\mu + \lambda + s)}$$

$$\therefore p^*(s) = \frac{\lambda}{\lambda + \mu}\left[\frac{1}{s} - \frac{1}{\lambda + \mu + s}\right] + \frac{1}{\mu + \lambda + s}$$

$$\therefore p^*(s) = \frac{\lambda}{\lambda + \mu}\cdot\frac{1}{s} + \frac{\mu}{\lambda + \mu}\cdot\frac{1}{\mu + \lambda + s}$$

$$\therefore p(t) = \frac{\lambda}{\lambda + \mu} + \frac{\mu}{\lambda + \mu}\,e^{-(\lambda + \mu)t} \qquad \text{as before.}$$

The expression for $p(t)$ gives the probability that the operator is busy at time t and this is the physical interpretation. However, we can go further. As $t \to \infty$, $e^{-(\lambda+\mu)t} \to 0$, so

$$\underset{t\to\infty}{\text{Limit}}\ p(t) = \frac{\lambda}{\lambda + \mu}$$

This represents the steady-state probability that the operator is busy after a long time. It also represents the long-run proportion of time for which the operator is busy. The long-run proportion of time for which the operator is free is

$$1 - \frac{\lambda}{\lambda + \mu} = \frac{\mu}{\lambda + \mu}$$

In fact the problem is identical, mathematically, to the very simple system with α replaced by λ and β by μ. A call arriving (at random when the operator is free at rate λ) corresponds to the machine breaking down (at random when it is running at rate α). The time to deal with the call corresponds to the time to repair the machine. $p(t)$, the probability that the operator is busy, corresponds to the probability that the machine is stopped and undergoing repair, $p_S(t)$. $1 - p(t)$, the probability that the operator is idle, corresponds to $p_R(t)$, the probability that the machine is running. The substitution of α for λ and β for μ confirms these results.

The modelling processes have differed but have arrived at the same result by different methods.

2.4 A queue with one server, random arrivals and negative exponential service

For a single-server queue we let $N(t)$ denote the number in the system at time t. $N(t)$ includes the customer who is currently being served (if there is one). Thus for the representation overleaf there are five customers in the system. There are four in the queue and one is being served (in the service box ⊠; see Fig. 2.3).

Figure 2.4 shows a possible realisation of a single-server queue. $N(t)$ denotes the number of customers in the system at time t. $N(t)$ increases by one when a new customer arrives and decreases by one when a customer's service terminates. A busy

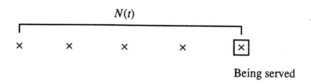

Fig. 2.3

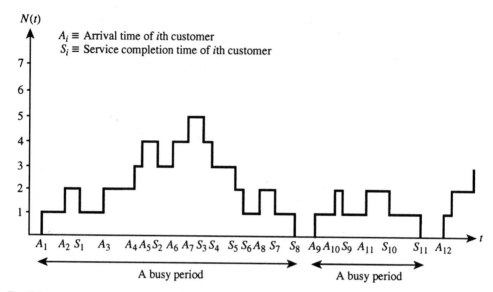

Fig. 2.4

period for the server starts when the server starts to serve a new customer prior to the system being empty, and ends when the system next becomes empty. Figure 2.4 illustrates two such periods.

This realisation has been constructed without any regard to the specific nature of the arrival pattern or the service time distribution. A shorthand notation has been constructed (by Professor D.G. Kendall) to describe queueing systems. It takes the form: inter-arrival time/service time/number of servers. Thus Fig. 2.4 is a realisation of a G/G/1 system, i.e. general inter-arrival time distribution, general service time distribution, one server. Another system, the D/G/2 system, would represent deterministic arrivals (at regular intervals), general service time distribution and two servers. The notation M/M/1 is used for the system having random arrivals, i.e. negative exponential inter-arrival time, negative exponential service time and one server. Here the M stands for Markovian.

If we consider the arrival pattern and service pattern from the example of the doctor's surgery in the Introduction, but suppose that the doctor was present from the outset, we can represent this on a similar diagram as an example of an M/M/1 queue. The arrival times of patients are: 11, 12, 14, 19, 21, 22, 30, 39, 45, 49, 67, 71, 78, 87, 115, 116 minutes. The service times are (in minutes): 2, 3, 10, 3, 2, 1, 5, 4, 11, 5, 3, 10, 7, 15, 5, 1.

We now have the arrival and service processes going on together, although we cannot serve a customer unless there is a customer present. Also the service of a customer cannot commence until the service of the previous customer has finished (see Table 2.1). This information can then be represented on a diagram showing $N(t)$ against t (Fig. 2.5).

Table 2.1

Patient number (i)	Arrival time (A_i)	Time into service	Service time	Departure time (S_i)
1	11	11	2	13
2	12	13	3	16
3	14	16	10	26
4	19	26	3	29
5	21	29	2	31
6	22	31	1	32
7	30	32	5	37
8	39	39	4	43
9	45	45	11	56
10	49	56	5	61
11	67	67	3	70
12	71	71	10	81
13	78	81	7	88
14	87	88	15	103
15	115	115	5	120
16	116	120	1	121

$$\text{Time into service} = \text{Max}\begin{cases} \text{Arrival time} \\ \text{Departure time of previous customer} \end{cases}$$

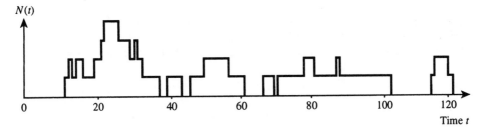

Fig. 2.5

So much for a painstaking hand simulation of 16 customers in an M/M/1 system. Let us now revert to mathematical modelling. We consider a typical small interval $(t, t + \delta t)$ and the activity that can take place in that interval. With random arrivals at rate α,

$$\alpha \delta t = \Pr(\text{arrival in } (t, t + \delta t))$$
$$1 - \alpha \delta t = \Pr(\text{no arrival in } (t, t + \delta t))$$

Inter-arrival times have a negative exponential distribution with p.d.f. $\alpha e^{-\alpha t}$; mean $1/\alpha$. With exponential service at rate β, the service time has a negative exponential distribution with probability density function $\beta e^{-\beta t}$; mean $1/\beta$.

$$\Pr(\text{ongoing service finishes in } (t, t + \delta t)) = \beta \delta t$$
$$\Pr(\text{ongoing service does not finish in } (t, t + \delta t)) = 1 - \beta \delta t$$

The above transition probabilities are correct to a term $o(\delta t)$. Let

$$\Pr(N(t) = n \,|\, N(0) = 0) = p_n(t)$$

Then to first order in δt

$$p_0(t + \delta t) = p_0(t)(1 - \alpha\delta t) + p_1(t)(1 - \alpha\delta t)\beta\delta t$$

There are none in the system at time $t + \delta t$ if there are none at t and no arrivals in $(t, t + \delta t)$ or one at t and a service completion and no arrival during δt.

$$p_1(t + \delta t) = p_0(t)\alpha\delta t + p_1(t)(1 - \alpha\delta t)(1 - \beta\delta t) + p_2(t)(1 - \alpha\delta t)\beta\delta t$$

There is one in the system at $t + \delta t$ if there are none at t and one arrival during δt, or one at t and no arrivals and no service completion during δt or two at t and no arrivals and one service completion during δt.

$$p_n(t + \delta t) = p_{n-1}(t)\alpha\delta t(1 - \beta\delta t) + p_n(t)(1 - \alpha\delta t)(1 - \beta\delta t)$$
$$+ p_{n+1}(t)(1 - \alpha\delta t)\beta\delta t$$

etc.

These three alternatives with appropriate changes in the value of n hold for $n \geq 1$. Thus after a little manipulation and as $\delta t \rightarrow 0$

$$\frac{dp_0(t)}{dt} = -\alpha p_0(t) + \beta p_1(t)$$

$$\frac{dp_1(t)}{dt} = \alpha p_0(t) - (\alpha + \beta)p_1(t) + \beta p_2(t) \qquad \text{etc.} \qquad (2.4)$$

$$\frac{dp_n(t)}{dt} = \alpha p_{n-1}(t) - (\alpha + \beta)p_n(t) + \beta p_{n+1}(t) \quad \text{etc.}$$

These differential equations are very hard to solve. The transient solution is difficult to find. However, if $\rho = \alpha/\beta < 1$ (ρ is the average number of new customers to arrive during a service time; the traffic intensity) there is a steady-state solution in which

$$\text{Limit}_{t \rightarrow \infty} p_n(t) \rightarrow p_n; \qquad \text{Limit}_{t \rightarrow \infty} \frac{dp_n(t)}{dt} = 0$$

$$\therefore \quad -\alpha p_0 + \beta p_1 = 0$$
$$\alpha p_0 - (\alpha + \beta)p_1 + \beta p_2 = 0 \qquad (2.5)$$
$$\alpha p_1 - (\alpha + \beta)p_2 + \beta p_3 = 0$$
$$\alpha p_{n-1} - (\alpha + \beta)p_n + \beta p_{n+1} = 0$$

with solution (obtained in succession)

$$p_1 = \frac{\alpha}{\beta} p_0 = \rho p_0$$

$$p_2 = \frac{\alpha}{\beta} p_1 = \rho^2 p_0$$

$$p_3 = \frac{\alpha}{\beta} p_2 = \rho^3 p_0$$

etc.

$$p_n = \rho^n p_0$$

and since $\sum_{n=0}^{\infty} p_n = 1$ we have

$$p_0(1 + \rho + \rho^2 + \rho^3 + \cdots) = 1$$

$$\therefore p_0 = (1 - \rho)$$

and so

$$p_n = (1 - \rho)\rho^n \qquad (2.6)$$

N.B. $\rho < 1$. If $\rho \geq 1$, a steady-state solution *does not* exist. The arrival rate must be strictly less than the service rate.

An alternative way to solve equations (2.5) is to use the generating function:

$$-\alpha p_0 + \beta p_1 = 0$$

$$\alpha p_0 - (\alpha + \beta)p_1 + \beta p_2 = 0$$

$$\alpha p_1 - (\alpha + \beta)p_2 + \beta p_3 = 0$$

$$\alpha p_2 - (\alpha + \beta)p_3 + \beta p_4 = 0$$

etc.

Let $P(z) = \sum_{n=0}^{\infty} p_n z^n$, multiply the equations above in turn by z^0, z^1, z^2, z^3, etc. and add. Then

$$\alpha z \sum_{n=0}^{\infty} p_n z^n - \alpha \sum_{n=0}^{\infty} p_n z^n - \beta \sum_{n=1}^{\infty} p_n z^n + \frac{\beta}{z} \sum_{n=1}^{\infty} p_n z^n = 0$$

$$\therefore \ \alpha z P(z) - \alpha P(z) - \beta(P(z) - p_0) + \frac{\beta}{z}(P(z) - p_0) = 0$$

$$\therefore \ P(z)[\alpha z^2 - (\alpha + \beta)z + \beta] = -\beta(z - 1)p_0$$

$$\therefore \ P(z) = \frac{-p_0(z - 1)}{\rho z^2 - (1 + \rho)z + 1} = \frac{-p_0(z - 1)}{(z - 1)(\rho z - 1)}$$

$$\therefore \ P(z) = \frac{p_0}{(1 - \rho z)}$$

Note that since $P(z)$ is regular for $|z| \leq 1$ the zero of the denominator at $z - 1 = 0$ must cancel with $z - 1$ in the numerator. The other zero of the denominator $z = 1/\rho$ is outside the unit circle if $\rho < 1$, which it must be.

$$\therefore \ P(z) = \frac{p_0}{1 - \rho z}$$

We can find p_0 since $P(1) = 1$. Thus $p_0 = (1 - \rho)$

$$P(z) = \frac{(1 - \rho)}{(1 - \rho z)}$$

$$= (1 - \rho)[1 + \rho z + \rho^2 z^2 + \rho^3 z^3 + \cdots] \qquad (2.7)$$

$$\therefore \ p_n = \text{coeff. of } z^n = (1 - \rho)\rho^n$$

Note:

1. p_n is the probability in the steady state that there are n customers in the system at an arbitrary moment;
2. p_n is the long-run proportion of time that there are n customers in the system.

Both of these interpretations are valid and useful.

Our modelling has resulted in equations (2.5) for the steady-state situation. We have solved these equations at (2.6).

2.5   Properties of the M/M/1 queueing system

$$\Pr(N = n) = p_n = (1 - \rho)\rho^n \qquad \text{where } \rho = \frac{\alpha}{\beta} < 1$$

The proportion of time the server is idle is

$$p_0 = (1 - \rho) \tag{2.8}$$

and the proportion of time the server is busy is

$$(1 - p_0) = \rho \tag{2.9}$$

The mean number in the system

$$E[N] = \sum_{n=0}^{\infty} np_n = (1 - \rho)\Sigma n\rho^n = \frac{(1 - \rho)\rho}{(1 - \rho)^2}$$

$$\left[\begin{array}{l} f(\rho) = \dfrac{1}{1 - \rho} = 1 + \rho + \rho^2 + \rho^3 + \cdots \\[3mm] f'(\rho) = \dfrac{1}{(1 - \rho)^2} = 1 + 2\rho + 3\rho^2 + \cdots \end{array} \right]$$

Therefore the mean number in the system, $E[N]$ is

$$E(N) = L = \frac{\rho}{1 - \rho} = \frac{\alpha}{\beta - \alpha} \tag{2.10}$$

We could of course use the generating function

$$P(z) = \frac{(1 - \rho)}{(1 - \rho z)}$$

$$P'(z) = \frac{\rho(1 - \rho)}{(1 - \rho z)^2}$$

$$E(N) = P'(1) = \frac{\rho}{1 - \rho} \qquad \text{as before.}$$

The mean number in the queue

$$L_q = \sum_{n=1}^{\infty} (n-1)p_n \qquad \text{(note the lower limit in the sum)}$$

$$= \sum_{n=1}^{\infty} np_n - \sum_{n=1}^{\infty} p_n = \sum_{n=1}^{\infty} np_n - (1-p_0) = L - \rho$$

$$\therefore L_q = \frac{\rho}{1-\rho} - \rho = \frac{\rho^2}{(1-\rho)} = \frac{\alpha^2}{\beta(\beta-\alpha)} \tag{2.11}$$

Also we have

$$L = L_q + \rho$$

Note that the number in the queue is not the number in the system minus one. This is only the case when $n \geq 1$. Note also that for ρ small, L and L_q are small but the proportion of time the server is idle is large (near to 1). On the other hand as ρ approaches 1 the server is busy but the average number in the system (or the queue) gets large.

Example 2.2

A repairman services and repairs domestic vacuum cleaners which are brought into his shop. Cleaners for repair arrive at random at an average rate of four per day (the repairman works an 8-hour day). The time to repair a cleaner varies with the type of repair needed, but has approximately a negative exponential distribution with mean 90 minutes. Find the proportion of time the repairman is idle and the average number of vacuum cleaners in the shop which are waiting to be repaired. For a typical cleaner brought in for repair, how many cleaners are there in front of it?

We take our time unit to be hours so that we have here a single-server queue with random arrivals at an average rate of $\alpha = 1/2$ a cleaner per hour (it may read rather strange that way but that is what it means) and $1/\beta = 3/2$, i.e. $\beta = 2/3$. Thus in our notation, $\rho = \alpha/\beta = 3/4$ and is less than 1. Therefore, $p_0 = 1 - \rho = 1/4$ and this represents the proportion of time that the repairman is not actually working on vacuum cleaners. To suggest he is idle is perhaps incorrect, since he can probably find other useful things to do.

The average number of vacuum cleaners waiting to be repaired is given by the average number in the queue L_q. By (2.11)

$$L_q = \frac{\rho^2}{1-\rho}$$

$$= \frac{9}{16} \bigg/ \frac{1}{4} = \frac{9}{4}$$

The number of cleaners in front of a cleaner just brought in is given by L, the average number in the system. This is given by $L = \rho/(1-\rho) = 3$.

The distribution of queueing time

Let X represent the queueing time, in the steady state, of an arbitrary customer. Let X have probability density function $\varphi(x)$

$$\Pr(x \leq X \leq x + \delta x) = \varphi(x)\,\delta x \qquad \text{to first order in } \delta x$$

$$= \Pr(n \text{ in the system on arrival, } and \ (n-1) \text{ services during } x,$$
$$and \text{ one service completion in } \delta x \text{ for } n = 1, 2, 3, \ldots)$$

We can represent 'what is going on' in diagrammatic form. On arrival our marked (but none the less arbitrary) customer, ⊗, finds n others (5 in this case) in the system. Our stop watch is then pressed, and shows zero. By time x on this watch, $(n-1)$ customers have been served so that our customer is at the head of the queue. Other customers may have arrived later. Then in $x + \delta x$, the customer being served (the one just in front of our customer when he arrived, *) has his/her service completed.

Time 0 on stop watch   ⊗ * × × × ⊠
Time x on stop watch   × × ⊗ ⊠
Time $x + \delta x$ on stop watch   × × ⊠ * →

$$\therefore \; \varphi(x)\,\delta(x) = \sum_{n=1}^{\infty} \frac{p_n(\beta x)^{n-1}\mathrm{e}^{-\beta x}}{(n-1)!}\, \beta\delta x \qquad \text{for } x > 0$$

$$\therefore \; \varphi(x) = \sum_{n=1}^{\infty} \frac{(1-\rho)\rho^n(\beta x)^{n-1}\mathrm{e}^{-\beta x}}{(n-1)!}\, \beta$$

$$= \left(1 - \frac{\alpha}{\beta}\right)\mathrm{e}^{-\beta x}\alpha \sum_{n=1}^{\infty} \frac{(\alpha x)^{n-1}}{(n-1)!}$$

$$\therefore \; \varphi(x) = \alpha\left(1 - \frac{\alpha}{\beta}\right)\mathrm{e}^{-(\beta-\alpha)x}; \qquad x > 0 \tag{2.12}$$

Note that

$$\int_0^{\infty} \varphi(x)\,\mathrm{d}x = \frac{\alpha}{\beta} \qquad (\neq 1)$$

For the *complete* distribution of queueing time we need to remember that if the system is empty on arrival the queueing time is zero.

$$\therefore \; \Pr(X = 0) = p_0 = 1 - \frac{\alpha}{\beta}$$

and

$$1 - \frac{\alpha}{\beta} + \alpha\left(1 - \frac{\alpha}{\beta}\right)\int_0^{\infty} \mathrm{e}^{-(\beta-\alpha)x}\,\mathrm{d}x = 1$$

The average time spent waiting in the queue is $E[X] = W_q$ and so

$$W_q = \alpha\left(1 - \frac{\alpha}{\beta}\right)\int_0^{\infty} x\mathrm{e}^{-(\beta-\alpha)x}\,\mathrm{d}x = \frac{\alpha\left(1 - \dfrac{\alpha}{\beta}\right)}{(\beta-\alpha)^2}$$

$$W_q = \frac{\alpha}{\beta(\beta-\alpha)} \tag{2.13}$$

Note $L_q = \alpha W_q$, the average number of customers who arrive during the queueing time of the previous customer.

If Y represents the time spent in the system by a customer, then

$$Y = \underset{\substack{\text{Queueing} \\ \text{time}}}{X} + \underset{\substack{\text{Service} \\ \text{time}}}{S}$$

where Y has p.d.f. $\psi(y) = (\beta - \alpha)e^{-(\beta-\alpha)y}$; $y \geq 0$ (see question 5 of Examples 5).

$$E[Y] = W = \frac{1}{\beta - \alpha}$$

Note

$$E[Y] = E[X] + E[S]$$

$$= \frac{\alpha}{\beta(\beta - \alpha)} + \frac{1}{\beta} = \frac{1}{\beta - \alpha} \tag{2.14}$$

and

$$L = \alpha W \tag{2.15}$$

The server's busy period

The server is either idle or busy. Idle periods and busy periods alternate through time.

Busy		Busy				Busy				
	Idle				Idle			Idle		

If we consider a long period of time of duration T, say, then the length of time the server is idle is $p_0 T = (1 - \rho)T$, and the length of time the server is busy is $(1 - p_0)T = \rho T$. The average length of an idle period is $1/\alpha$. Such periods, the time until the next arrival, are random variables with p.d.f. $\alpha e^{-\alpha t}$ (refer back to the Introduction). Thus the expected number of such idle periods is $(1 - \rho)T\alpha$, which equals the expected number of busy periods. Thus the average length of a busy period is

$$\frac{\rho T}{(1 - \rho)T\alpha} = \frac{\rho}{(1 - \rho)\alpha} \tag{2.16}$$

Example 2.3

Cars to a customer car-wash arrive at random at an average rate of 5 per hour. The time for a car to be washed by its owner has a negative exponential distribution with mean 6 minutes. Find:

(a) the average number of cars in and at the wash;
(b) the proportion of cars who on arrival proceed to the wash without any delay;
(c) the average time a car spends waiting before its wash commences (can you verify this result using your answer to (a)?);
(d) the average length of time that the wash is in continuous action.

Here we have a single-server queue with random arrivals at rate $\alpha = 1/12$, and negative exponential service at rate $\beta = 1/6$, where the time unit used is 1 minute. Thus,

$$\rho = \frac{\alpha}{\beta} = \frac{1}{2}$$

For

(a) we need L, the average number in the system, and this is given by (2.10)

$$L = \tfrac{1}{2}\bigg/\big(1 - \tfrac{1}{2}\big) = 1$$

(b) Such cars find an empty system on arrival. The proportion of such cars is thus

$$p_0 = 1 - \rho = \tfrac{1}{2}$$

(c) The time required is the average time spent waiting in the queue (not in the system, which would include the wash-time). In this case by (2.13)

$$W_q = \frac{\alpha}{\beta(\beta - \alpha)} = \frac{\tfrac{1}{12}}{\tfrac{1}{6}\big(\tfrac{1}{6} - \tfrac{1}{12}\big)} = 6 \text{ minutes}$$

From (a) a typical car arriving at the wash will find one $(L = 1)$ other car in the system in front of it. The wash-time (or residual wash-time, see Examples 1, question 7) of this car will have a negative exponential distribution with mean 6 minutes. This mean value is the mean waiting time as given.

(d) The average length of time is given by the average length of a busy period, from the time a customer arrives at an empty wash to the time the wash is next empty. On using (2.16) we obtain the result

$$\frac{\rho}{(1 - \rho)\alpha} = \frac{12 \times \tfrac{1}{2}}{1 - \tfrac{1}{2}} = 12 \text{ minutes}$$

2.6 Some general considerations

From the solution (2.6) of equations (2.5) we have been able to obtain closed-form mathematical expressions for a number of important quantities connected with the M/M/1 system. If we know α and β (we can then find ρ) the expressions (2.8)–(2.16) can easily be evaluated. These allow us to predict quantities such as the proportion of time the server is idle, the mean number in the queue, the mean time a customer spends in the system, etc. without going through the tedious process of simulating the behaviour of the system over a very long period.

The latter does involve a great deal of work. We have simulated the doctor's surgery for a 2-hour period. Imagine the work involved in doing this for several thousands of patients. The mathematical modelling, when it is successful, has great merit.

Some of the mathematical results deserve more careful consideration. In the first place, the steady-state solution only exists if $\rho < 1$ (strictly less than 1). At one time it was felt that the way to run a queueing system was to balance the arrival and service rates (make $\alpha = \beta$ and $\rho = 1$). Of course if we do this the number in the system will just grow without limit. There is only a steady-state solution if $\rho < 1$. A queue is *not* a symmetric system, Customers who have not arrived cannot be served (see question 1 of Examples 5).

If ρ is small the server is idle for a large proportion of the time, whereas if ρ is near to 1 the server is kept busy for a large proportion of the time [(2.8) and (2.9)]. However, it is only when ρ is small that the mean number in the system, the mean number in the queue, the average waiting time and time in the system are reasonably small. For ρ near to 1 (i.e. α and β are nearly the same) the presence of $1-\rho$ (or $\beta - \alpha$) in the denominator in (2.10), (2.11), (2.13) and (2.14) causes the above quantities to increase very rapidly indeed. If we keep the server busy we must inevitably keep a large number of customers waiting for a long time. If we keep our customers happy with only a few waiting for a short time, we must accept that our server will not be fully utilised.

There have been recent suggestions in connection with the National Health Service, for example, that expensive equipment (operating theatres, scanners, etc.) should be used efficiently (does that mean fully utilised?) whilst at the same time waiting lists and waiting times should be reduced. If the M/M/1 model is any sort of reasonable model for such systems it suggests that it is quite impossible to achieve all of these goals simultaneously. If we want to cut waiting lists and waiting times we must supply adequate service facilities which will mean they are not fully utilised. That is the way things are!

2.7 Examples 5

1 Consider the following three examples of the D/D/1 system.

(i) Customers to a single-server queue arrive at 5-minute intervals after time 0 when the system opens. The server takes exactly 10 minutes to serve a customer. Draw a graph of $N(t)$ against t where $N(t)$ is the number in the system at time t.

For the customer who arrives at time 105 minutes, find the number in the system in front of that customer on arrival and the time that customer waits for service.

Repeat the above for the customer who arrives at time 110 minutes. For what proportion of time is the server idle?

(ii) Customers to a single-server queue arrive at 10-minute intervals after the system opens and take exactly 10 minutes to be served. Draw a graph of $N(t)$ against t.

For a customer, find the number in the system on arrival, the time spent waiting for service, and the time spent in the system. For what proportion of time is the server idle?

(iii) Customers to a single-server queue arrive at 10-minute intervals after the system opens and take exactly 5 minutes to be served. Draw a graph of $N(t)$ against t.

For a customer, find the number in the system on arrival, the time spent waiting for service, and the time spent in the system. For what proportion of time is the server idle?

Compare the results (iii) above with the corresponding outcomes for the steady-state solution of an M/M/1 system.

2 For the M/M/1 system (equations (2.5) and (2.6)) use the probability generating function (2.7) to show that

$$\text{Var}[N] = \frac{\rho}{(1-\rho)^2}$$

3 Lorries arrive at a warehouse at random at an average rate of 1 every 50 minutes. The time to unload a lorry has an exponential distribution with mean 40 minutes. Find the proportion of time that the unloading bay is free and find the mean number of lorries idle (i.e. in the system waiting or being unloaded).

If new unloading equipment were installed the mean unloading time would be reduced to 30 minutes. The cost of the new equipment would be £600/week above present costs. If a lorry costs £10/hour when it is idle and works for 40 hours/week, does it seem sensible to install the new equipment?

4 Barges arrive at a lock at an average rate of 4/hour. The time to pass through the lock has a mean of 10 minutes.

What is the mean time a barge has to wait until the lock is free? What is the probability that more than three barges are waiting? What assumptions have you made?

5 If $Y = X + S$ where X has p.d.f. $\varphi(x) = \alpha(1 - \alpha/\beta)e^{-(\beta-\alpha)x}$; $x > 0$,
$\Pr(X = 0) = 1 - \alpha/\beta$, and S has p.d.f. $b(s) = \beta e^{-\beta s}$, $s \geq 0$ show that Y has p.d.f.

$$\psi(y) = (\beta - \alpha)e^{-(\beta-\alpha)y}; \qquad y \geq 0$$

Find $E[Y]$ [see equation (2.14)].

6 At what average rate should a cashier in a supermarket work to ensure that 90% of customers do not have to wait in the queue for more than 4 minutes? There is only one cashier, customers arrive at random at a rate of 12/hour and service time has a negative exponential distribution.

2.8 Solutions to Examples 5

1 (i) The arrivals come faster than the server can serve them and so the number in the system steadily increases, as Fig. 2.6 shows.

By 105 minutes, 21 customers have arrived (including the one at 105) and 10 customers have been served. Thus this customer finds 10 in the system and has to

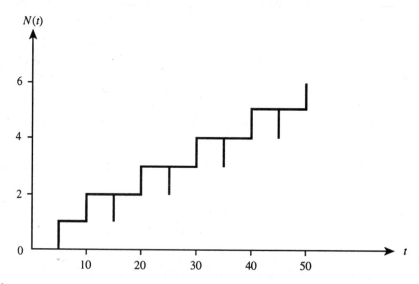

Fig. 2.6

wait for 100 minutes for his service to begin. By 110 minutes, 22 customers have arrived (including the one at 110) and 10 have been served. Thus this customer finds 11 in the system but, since the customer currently being served only has 5 minutes of service to run, the waiting time is 105 minutes. The server has no idle time after the first 5 minutes.

(ii) Here the system is in exact balance as Fig. 2.7 shows. As the service of one customer ends, so the next customer arrives.

Each customer finds an empty system (just) on arrival and so does not have to wait for service and spends 10 minutes in the system. The server is kept fully occupied and so has no idle time after the first customer has arrived.

(iii) Once again the behaviour of this deterministic system can be simply shown (Fig. 2.8). Each customer finds an empty system, going straight into service without queueing and spends 5 minutes in the system. The server is idle for half the time with 5-minute busy periods alternating with 5-minute idle periods.

If we compare (iii) to an M/M/1 system with $\rho = \alpha/\beta = (1/10)/(1/5) = 1/2$ then from (2.10), (2.13), (2.14), (2.9), $L = 1$, $W_q = 5$ and $W = 10$ whilst the proportion of time for which the server is idle is $1/2$.

The reason that there is congestion in the M/M/1 system, and not the D/D/1 system in this case, is because in the former system the random nature of the arrivals will sometimes mean that in the short run there will be temporary periods in which more customers than usual arrive and the server will not be able to deal with them all without a queue forming. This does not happen in the D/D/1 system. For both systems the server is busy for half the time.

Of course periods when the service mechanism performs better than average will also feature and will help to relieve the congestion, but these cannot occur unless the congestion to be relieved is already there. Although a steady state exists, the short run irregularities still lead to queueing and waiting.

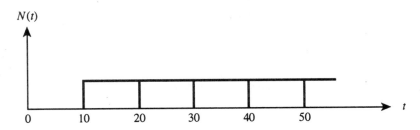

Fig. 2.7

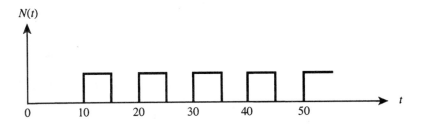

Fig. 2.8

In the situation described by (ii), there is not even a steady state for the M/M/1 system (see also the Introduction). The arrival peaks will cause congestion which the server will not be able to eliminate in the long run. As has been said before, a queue is not symmetric. Customers can only be served after they have arrived, not before the queue has developed.

2

$$P(z) = \frac{(1-\rho)}{(1-\rho z)}; \qquad P'(z) = \frac{\rho(1-\rho)}{(1-\rho z)^2}; \qquad P''(z) = \frac{2\rho^2(1-\rho)}{(1-\rho z)^3}$$

$$\therefore \ P'(1) = E[N] = \frac{\rho(1-\rho)}{(1-\rho)^2} = \frac{\rho}{(1-\rho)}$$

$$P''(1) = E[N(N-1)] = \frac{2\rho^2(1-\rho)}{(1-\rho)^3} = \frac{2\rho^2}{(1-\rho)^2}$$

$$\therefore \ E[N^2] = \frac{2\rho^2}{(1-\rho)^2} + \frac{\rho}{(1-\rho)}$$

$$\therefore \ \mathrm{Var}[N] = E[N^2] - \{E[N]\}^2$$

$$= \frac{2\rho^2}{(1-\rho)^2} + \frac{\rho}{(1-\rho)} - \frac{\rho^2}{(1-\rho)^2} = \frac{\rho}{(1-\rho)^2}$$

3 In the notation of (2.5) and sequel, for the first situation (before new equipment is installed) $\alpha = 1/50$, $\beta = 1/40$ and $\rho = 4/5$ ($\alpha = 1/50$ of a lorry per *minute*, $\beta = 1/40$ of a service per *minute*, so time is measured in minutes).

$$\therefore \ p_0 = 1 - \rho = \tfrac{1}{5} = 0.2; \qquad L = E[N] = \frac{\rho}{1-\rho} = 4$$

After the change $\alpha' = 1/50$, $\beta' = 1/30$, $\rho' = 3/5$ and $p_0 = 0.4$ and $L = 3/2$. Thus the average number of idle lorries in the system is reduced by 2.5. This saves $2.5 \times 40 \times 10 = £1000/\text{week}$. Since cost of new equipment is only £600/week, it is good to install it.

4 Again with the earlier notation $\alpha = 4$, $\beta = 6$ (per hour), $\rho = \alpha/\beta = 2/3$

$$\therefore \ E[X] = W_q = \frac{\alpha}{\beta(\beta - \alpha)} = \frac{4}{6(6-4)} = \tfrac{1}{3} \ \text{hour}$$

Pr(more than 3 barges are waiting) = Pr(more than 4 barges in the system)

$$= \sum_{n=5}^{\infty} p_n = (1-\rho)\rho^5[1 + \rho + \rho^2 + \cdots]$$

$$= \rho^5 = \left(\tfrac{2}{3}\right)^5$$

We have assumed random arrivals and exponential service.

5 $Y = X + S$. When X is zero (with probability $1 - \alpha/\beta$), then $Y = S$.

$$\therefore f(y) = \left(1 - \frac{\alpha}{\beta}\right)\beta e^{-\beta y} + \int_0^y \frac{\alpha}{\beta}(\beta - \alpha)e^{-(\beta - \alpha)x}\beta e^{-\beta(y - x)}\, dx$$

$$= (\beta - \alpha)e^{-\beta y} + \alpha(\beta - \alpha)e^{-\beta y}\int_0^y e^{\alpha x}\, dx$$

$$= (\beta - \alpha)e^{-\beta y} + (\beta - \alpha)e^{-\beta y}[e^{\alpha y} - 1] = (\beta - \alpha)^{-(\beta - \alpha)y}; \qquad y \geqslant 0$$

Thus Y has a negative exponential distribution with parameter $(\beta - \alpha)$ so that $E[Y] = 1/(\beta - \alpha)$.

6 We make the assumption of random arrivals at rate $\alpha = 1/5$, and negative exponential service at rate β (to be found). Then we can use the results (2.8) and sequel.

With minutes as the time units, $\alpha = 1/5$, β is unknown. X, the time spent in the queue, has p.d.f.

$$\frac{\alpha(\beta - \alpha)e^{-(\beta - \alpha)x}}{\beta} \qquad \text{for } x > 0$$

Our condition to be satisfied is

$$\int_4^\infty \frac{\alpha(\beta - \alpha)}{\beta} e^{-(\beta - \alpha)x}\, dx \leqslant 0.1$$

Less than 10% queue for more than 4 minutes.

An alternative but equivalent condition is that $\Pr[X \leqslant 4] \geqslant 0.9$, i.e.

$$1 - \frac{\alpha}{\beta} + \frac{\alpha}{\beta}\int_0^4 (\beta - \alpha)e^{-(\beta - \alpha)x}\, dx \geqslant 0.9$$

$$\therefore \left[-\frac{\alpha}{\beta} e^{-(\beta - \alpha)x}\right]_4^\infty \leqslant 0.1$$

$$\therefore \frac{1}{5\beta} e^{-(\beta - 1/5)4} \leqslant 0.1$$

$$\therefore \frac{2}{\beta} e^{-4\beta}e^{4/5} \leqslant 1$$

$$\therefore \beta e^{4\beta} \geqslant 2e^{4/5}$$

The solution of the equation $\beta e^{4\beta} = 2e^{4/5}$ has to be obtained numerically and has value 0.5315. Thus β, the rate of service, must exceed 0.5315. We can also say that $1/\beta$, the mean service time, must be less than $1/0.5315 = 1.88$ minutes.

3
Birth–death models

We can extend the previous model to the case of random arrivals at rate α_n when there are n in the system and exponential service at rate β_n when there are n in the system. Thus the arrival and service rates depend on the number in the system (population, arrival $\equiv$ birth; service completion $\equiv$ death).

$p_n(t)$ is the probability there are n in the system at time t

$$\Pr(N(t) = n) = p_n(t)$$

Then arguing as before for the interval $(t, t + \delta t)$, to first order in δt;

$$p_0(t + \delta t) = p_0(t)(1 - \alpha_0 \delta t) + p_1(t)\beta_1 \delta t$$

$$p_1(t + \delta t) = p_0(t)\alpha_0 \delta t + p_1(t)(1 - \alpha_1 \delta t)(1 - \beta_1 \delta t) + p_2(t)\beta_2 \delta t$$

$$p_n(t + \delta t) = p_{n-1}(t)\alpha_{n-1}\delta t + p_n(t)(1 - \alpha_n \delta t)(1 - \beta_n \delta t) + p_{n+1}(t)\beta_{n+1}\delta t$$

There are n in the system at $t + \delta t$ if there are $(n-1)$ at time t, and an arrival and no service completion in δt or n at time t, and no arrival and no service completion in δt or $(n+1)$ at time t and no arrival but one service completion in δt.

The first equation is different because we cannot have -1 in the system and then an arrival (as we have said before a queue is not a symmetric system). Of course the arrival rates and service rates depend on n and this has been taken into account.

$$\therefore \quad \frac{\mathrm{d}p_0(t)}{\mathrm{d}t} = -\alpha_0 p_0(t) + \beta_1 p_1(t)$$

$$\frac{\mathrm{d}p_1(t)}{\mathrm{d}t} = \alpha_0 p_0(t) - (\alpha_1 + \beta_1)p_1(t) + \beta_2 p_2(t) \tag{3.1}$$

$$\frac{\mathrm{d}p_n(t)}{\mathrm{d}t} = \alpha_{n-1}p_{n-1}(t) - (\alpha_n + \beta_n)p_n(t) + \beta_{n+1}p_{n+1}(t) \qquad n \geqslant 1$$

etc.

In general it is not possible to find the time-dependent solution except in a few special cases. Further discussion will follow in Chapter 8. If a steady-state solution exists $p_n(t) \to p_n$ as $t \to \infty$ and $\mathrm{d}p_n(t)/\mathrm{d}t \to 0$ as $t \to \infty$ then the equations for the p_n are:

$$-\alpha_0 p_0 + \beta_1 p_1 = 0$$

$$\alpha_0 p_0 - (\alpha_1 + \beta_1)p_1 + \beta_2 p_2 = 0$$

$$\alpha_1 p_1 - (\alpha_2 + \beta_2)p_2 + \beta_3 p_3 = 0 \tag{3.2}$$

etc.

We can solve these in turn to obtain

$$p_1 = \frac{\alpha_0}{\beta_1} p_0, \qquad p_2 = \frac{\alpha_1}{\beta_2} p_1 = \frac{\alpha_0 \alpha_1}{\beta_1 \beta_2} p_0 \text{ etc.}$$

$$p_n = \frac{\alpha_0 \alpha_1 \alpha_2 \ldots \alpha_{n-1}}{\beta_1 \beta_2 \ldots \beta_n} p_0 \qquad (3.3)$$

and since $\sum_{n=0}^{\infty} p_n = 1$

$$p_0 = 1 \Big/ \left\{ 1 + \frac{\alpha_0}{\beta_1} + \frac{\alpha_0 \alpha_1}{\beta_1 \beta_2} + \frac{\alpha_0 \alpha_1 \alpha_2}{\beta_1 \beta_2 \beta_3} + \cdots \right\} \qquad (3.4)$$

A necessary condition for the existence of a steady-state solution is that $\sum_{n=0}^{\infty} (\alpha_0 \alpha_1 \ldots \alpha_{n-1})/(\beta_1 \beta_2 \ldots \beta_n)$ is convergent so that p_0 is non-zero.

Thus from (3.3) p_n is non-zero for finite values of n. Otherwise no steady state exists and in the long run the number in the system becomes infinite.

3.1 Special cases

The M/M/1 queue with limited waiting room

A queue with one server, random arrivals at rate α, exponential service at rate β and room for N customers. Then

$$\alpha_0 = \alpha_1 = \alpha_2 \ldots = \alpha_{N-1} = \alpha; \qquad \alpha_N = \alpha_{N+1} \ldots = 0$$
$$\beta_1 = \beta_2 \quad = \beta_N = \beta$$

Again let $\rho = \alpha/\beta$, then

$$p_n = p_0 \rho^n \qquad \text{for } n = 0, 1, 2, \ldots, N$$
$$p_n = 0 \qquad \text{for } n > N$$

Since $\sum_{r=0}^{N} p_r = 1$, $p_0 \sum_{r=0}^{N} \rho^r = 1$

$$\therefore \ p_0 = \frac{(1 - \rho)}{(1 - \rho^{N+1})}$$

$$p_n = \frac{\rho^n (1 - \rho)}{(1 - \rho^{N+1})}$$

Note that this result is true for any value of ρ. $\sum_{r=0}^{N} \rho^r$, which corresponds to (3.4), is finite for all values of ρ. Of course if $\rho < 1$ and we let $N \to \infty$, we get $p_n = (1 - \rho)\rho^n$ as in the M/M/1 system (with infinite waiting room).

p_N is the proportion of time that the system is full and is hence the proportion of *potential* customers who go elsewhere.

A model where customers are discouraged by a long queue

We assume one server, random arrivals (rate α), exponential service (rate β) but when there are n in the system only a proportion $1/(n+1)$ of the arrivals actually join the queue. The others go elsewhere. Thus $\alpha_n = \alpha/(n+1)$, $\beta_n = \beta$ for all n.

$$\therefore \ p_1 = p_0\frac{\alpha}{\beta}, \qquad p_2 = p_0\frac{\alpha^2}{\beta.2\beta} = \frac{1}{2!}\left(\frac{\alpha}{\beta}\right)^2 p_0, \qquad p_3 = \frac{1}{3!}\left(\frac{\alpha}{\beta}\right)^3 p_0 \text{ etc.}$$

$$p_n = \frac{\rho^n p_0}{n!} \qquad \text{where } \rho = \frac{\alpha}{\beta}$$

Since $\sum_{n=0}^{\infty} p_n = 1$

$$p_0\left[1 + \rho + \frac{\rho^2}{2!} + \frac{\rho^3}{3!} + \cdots\right] = 1$$

$$\therefore \ p_0 = e^{-\rho} \text{ and } p_n = \frac{\rho^n e^{-\rho}}{n!}$$

If N represents the number of customers in the system (in the steady state) then N has a Poisson distribution with parameter ρ.

$$\mathrm{E}[N] = \rho, \qquad \mathrm{Var}[N] = \rho$$

The proportion of time the server is free $= p_0 = e^{-\rho}$.

A queue with self-service. The M/M/∞ queue

$\alpha_n = \alpha$ and $\beta_n = n\beta$ (customers serve themselves, exponential at rate β).

$$\therefore \ p_1 = \frac{\alpha}{\beta}p_0, \qquad p_2 = \frac{\alpha^2}{2!\beta^2}p_0, \qquad p_3 = \frac{1}{3!}\left(\frac{\alpha}{\beta}\right)^3 p_0 \text{ etc.}$$

$$\therefore \ p_n = \frac{\rho^n e^{-\rho}}{n!}$$

where $\rho = \alpha/\beta$ as in the previous model.

In both of the previous models a steady-state solution exists for all values of ρ. The series

$$1 + \rho + \frac{\rho^2}{2!} + \frac{\rho^3}{3!} + \cdots$$

is always convergent with sum e^{ρ}.

Example 3.1

A tool rental shop has four floor sanders which are available for hire. Customers for floor sanders arrive at random at an average rate of one customer every 2 days. The rental time has a negative exponential distribution with mean 2 days. If the shop has no floor sanders available for hire, the customers go elsewhere and their trade is lost.

(i)   Find the proportion of potential trade that is lost.
(ii) Estimate the mean number of floor sanders rented out.
(iii) One of the sanders develops a fault and has to be repaired. Estimate the effect that this has on the proportion of trade that is lost.

We have a system in which the number of customers is equal to the number of sanders rented out. We can describe this system using the birth–death equations with

$$\alpha_0 = \alpha_1 = \alpha_2 = \alpha_3 = \tfrac{1}{2}; \qquad \alpha_4 = \alpha_5 = \cdots = 0$$

$$\beta_1 = \tfrac{1}{2}, \qquad \beta_2 = 2 \times \tfrac{1}{2}, \qquad \beta_3 = 3 \times \tfrac{1}{2}, \qquad \beta_4 = 4 \times \tfrac{1}{2}$$

(It is essentially a self-service system with up to four customers allowed in the system.) Thus on using (3.3) and the notation used in deriving this result

$$p_1 = p_0$$
$$p_2 = \tfrac{1}{2} p_1 = \tfrac{1}{2} p_0$$
$$p_3 = \tfrac{1}{3} p_2 = \tfrac{1}{6} p_0$$
$$p_4 = \tfrac{1}{4} p_3 = \tfrac{1}{24} p_0$$

Thus

$$p_0[1 + 1 + \tfrac{1}{2} + \tfrac{1}{6} + \tfrac{1}{24}] = p_0[\tfrac{65}{24}] = 1$$

$$\therefore p_0 = \tfrac{24}{65}, \; p_1 = \tfrac{24}{65}, \; p_2 = \tfrac{12}{65}, \; p_3 = \tfrac{4}{65} \text{ and } p_4 = \tfrac{1}{65}$$

Thus

(i) The proportion of potential trade lost $= p_4 = 1/65 \simeq 1.54\%$.
(ii) Mean number of sanders hired out

$$= 1 \times \tfrac{24}{65} + 2 \times \tfrac{12}{65} + 3 \times \tfrac{4}{65} + 4 \times \tfrac{1}{65} = \tfrac{64}{65}$$

(iii) If only three sanders are available then $\alpha_3 = 0$ and β_4 is no longer valid. Then

$$p_0'[1 + 1 + \tfrac{1}{2} + \tfrac{1}{6}] = 1$$

so that

$$p_0' = \tfrac{6}{16}, \; p_1' = \tfrac{6}{16}, \; p_2' = \tfrac{3}{16} \text{ and } p_3' = \tfrac{1}{16}$$

The proportion of trade lost increases to $1/16 = 6.25\%$.

Example 3.2

Customer impatience is modelled as follows for a particular system. Potential customers arrive at the single service point at random at rate α but only a proportion $e^{-\gamma n}$ of such customers actually join the system when they find n in the system on their arrival. Here γ is a positive constant. Service times have a negative exponential distribution with mean $1/\beta$. Find an expression for the probability that there are n customers in the system in the steady-state situation.

Here in our notation is $\alpha_n = \alpha e^{-\gamma n}$, $\beta_n = \beta$. Thus,

$$p_n = p_0 \frac{\alpha^n e^{-\gamma(1 + 2 + 3 + \cdots + n - 1)}}{\beta^n}$$

$$= p_0 \rho^n e^{\frac{-\gamma n(n-1)}{2}}$$

The value of p_0 is determined from

$$p_0\left\{ 1 + \sum_{n=1}^{\infty} \rho^n e^{\frac{-\gamma n(n-1)}{2}} \right\} = 1$$

The series in the brackets will converge very rapidly, but for given values of ρ and γ it is not possible to obtain a closed-form expression for the sum, and in a practical problem we will have to content ourselves with an approximate numerical value.

Example 3.3

A secretary provides a word-processing service for postgraduate theses and dissertations. Because this extra work is done in her own time she strictly controls her workload from this source. She will not accept a thesis if, when it is submitted, there are already, in addition to the thesis she is typing, two theses waiting to be typed.

A reasonable model for the system is, that requests to type a thesis arrive at random at an average rate of one every 60 days. The time to word-process a thesis has a negative exponential distribution with a mean of 30 days.

(i)   Find the average total workload in the system at any time.
(ii) Find the length of time an accepted thesis can expect to wait between submission and the commencement of typing.
(iii) How long can a student expect to wait from the time the thesis is submitted for typing until completion, provided of course the thesis is accepted?

(i) It is convenient to take the mean service time (30 days) as the unit of time. Thus we have a single-server queueing system with average arrival rate $\alpha = 1/2$ and service rate $\beta = 1$ with this time unit. No more than three customers (theses) are allowed in the system at any one time. Thus if p_n denotes the steady-state probability that there are n theses in the system

$$p_1 = \tfrac{1}{2}p_0, \ p_2 = \tfrac{1}{2}p_1 = \tfrac{1}{4}p_0 \text{ and } p_3 = \tfrac{1}{2}p_2 = \tfrac{1}{8}p_0$$

Thus

$$p_0\{1 + \tfrac{1}{2} + \tfrac{1}{4} + \tfrac{1}{8}\} = 1$$

$$\therefore p_0 = \tfrac{8}{15}, \ p_1 = \tfrac{4}{15}, \ p_2 = \tfrac{2}{15} \text{ and } p_3 = \tfrac{1}{15}$$

The average total workload is the average number of theses in the system and is given by

$$p_1 + 2p_2 + 3p_3 = \tfrac{4}{15} + \tfrac{4}{15} + \tfrac{3}{15} = \tfrac{11}{15}$$

(ii) We need to find the average time spent in the queue by a thesis which is accepted. Since $p_3 = 1/15$ only $14/15$ of the submitted theses are accepted. Thus if q_n denotes the probability that an accepted thesis finds n theses in the system on arrival

$$q_0 = p_0/\tfrac{14}{15}, \ q_1 = p_1/\tfrac{14}{15}, \ q_2 = p_2/\tfrac{14}{15}$$

$$\therefore q_0 = \tfrac{4}{7}, \ q_1 = \tfrac{2}{7}, \ q_2 = \tfrac{1}{7}$$

Thus the average number in the system found by an accepted thesis is $1 \times (2/7) + 2 \times (1/7) = 4/7$. The typing time or residual typing time (for the one currently being typed) has mean 1 unit (30 days). Thus the expected time spent in the queue by an accepted thesis is $4/7 \times 30 = 120/7$ days.

As an alternative we can repeat the argument which led to (2.12) for the distribution of the queueing time, which we denote by the random variable X. For an accepted thesis $\Pr(X = 0) = q_0 = 4/7$.

For $x > 0$, X has probability density function $\varphi(x)$ and

$$\varphi(x)\,\mathrm{d}x = \Pr(1 \text{ in system on arrival}) \times \Pr(0 \text{ completions in } (0,\,x))$$
$$\times \Pr(1 \text{ completion in } (x,\,x + \mathrm{d}x))$$
$$+ \Pr(2 \text{ in system on arrival}) \times \Pr(1 \text{ completion in } (0,\,x))$$
$$\times \Pr(1 \text{ completion in } (x,\,x + \mathrm{d}x))$$

Thus since $\beta = 1$

$$\varphi(x) = q_1 e^{-x} + q_2 x e^{-x}$$

where we use q_1 and q_2 (rather than p_1 and p_2) since we must condition on the fact that there must be fewer than three in the system for the arriving thesis to be accepted for typing.

Thus $\varphi(x) = 2/7(e^{-x}) + 1/7(xe^{-x})$ and bearing in mind that $\int_0^\infty x^n e^{-x}\,\mathrm{d}x = n!$ we see that

$$\Pr(X = 0) + \int_{0+}^\infty \varphi(x)\,\mathrm{d}x = \frac{4}{7} + \frac{3}{7} = 1$$

$$\mathrm{E}[X] = \int_{0+}^\infty x\varphi(x)\,\mathrm{d}x = \frac{2}{7}\int_{0+}^\infty xe^{-x}\,\mathrm{d}x + \frac{1}{7}\int_{0+}^\infty x^2 e^{-x}\,\mathrm{d}x$$

$$= \frac{4}{7} \text{ units}$$

Thus, as before, expected queueing time $= 120/7$ days.

(iii) To find the expected time in the system we simply add to this quantity the mean time to type or word-process the thesis, namely 30 days, so that the expected total time in the system (expected response time) is $330/7$ days, or nearly 7 weeks.

3.2 Examples 6

1 A service station has one pump and space for three cars on the forecourt. Potential customers in the form of cars needing petrol etc. pass at random at an average rate of one every 8 minutes. The time to serve each customer has an exponential distribution with mean 4 minutes.

Find the proportion of time that there are 0, 1, 2, 3 cars at the station. What proportion of potential custom is lost because of the limited space?

2 The owner of the station in question 1 can rent some land at the side which will give space for one extra car. This will cost £10/week. The mean profit per customer is £0.50. The station is open 10 hours/day. Do you think he should rent the land?

3 A car-park has room for N cars. They arrive at random at rate α and parking times have an exponential distribution with p.d.f. $\beta e^{-\beta t}$; $t \geqslant 0$. Calculate

(i) the proportion of time the park is empty;
(ii) the proportion of cars who cannot find a place.

You should assume that when the car-park is full, arriving cars go elsewhere immediately.

3.3   Solutions to Examples 6

1 In the notation of the notes $\alpha_0 = \alpha_1 = \alpha_2 = 1/8$; $\alpha_3 = \alpha_4 = \cdots = 0$;   and $\beta_1 = \beta_2 = \beta_3 = 1/4$.

$$p_1 = \tfrac{1}{2}p_0,\ p_2 = \tfrac{1}{4}p_0,\ p_3 = \tfrac{1}{8}p_0$$

Thus,

$$p_0(1 + \tfrac{1}{2} + \tfrac{1}{4} + \tfrac{1}{8}) = 1$$

$$\therefore p_0 = \tfrac{8}{15} \qquad \text{and} \qquad p_1 = \tfrac{4}{15},\ p_2 = \tfrac{2}{15},\ p_3 = \tfrac{1}{15}$$

Therefore $1/15$ ($=p_3$) of potential customers are lost (i.e. drive on since there is no room).

2 As above, but now $\alpha_3 = 1/8$ and $\beta_4 = 1/4$. Thus the probabilities for 0, 1, 2, 3, 4 in the system are by a virtually identical calculation

$$p_0 = \tfrac{16}{31},\ p_1 = \tfrac{8}{31},\ p_2 = \tfrac{4}{31},\ p_3 = \tfrac{2}{31},\ p_4 = \tfrac{1}{31}$$

In question 1, 14/15 of passing trade is realised. Profit each week = 14/15 × 1/8 × 600 × 7 × 0.5 = £245.

In question 2, 30/31 of passing trade is realised. Profit each week = 30/31 × 1/8 × 600 × 7 × 0.5 − 10 = £244.03.

Thus he does not gain sufficient to cover the rent and he should not rent the land.

3 $\alpha_0 = \alpha_1 \ldots = \alpha_{N-1} = \alpha;\ \alpha_N = \alpha_{N+1} = \cdots = 0;\ \beta_n = n\beta$ for $n = 1, \ldots, N$. $p_0, p_1 = \rho p_0$, $p_2 = \rho^2 p_0/2! \ldots p_N = \rho^N p_0/N!$ where $\rho = \alpha/\beta$.

$$\therefore\ p_0 = 1 \bigg/ \left\{ 1 + \rho + \frac{\rho^2}{2!} + \cdots + \frac{\rho^N}{N!} \right\} \qquad \text{This is the answer to (i)}$$

$$p_N = \frac{\rho^N}{N!} \bigg/ \left\{ 1 + \rho + \frac{\rho^2}{2!} + \cdots + \frac{\rho^N}{N!} \right\} \qquad \text{This is the answer to (ii)}$$

3.4   A two-server queue – the M/M/2 queue

Random arrivals at rate α, negative exponential service at rate β from each of two servers. We can treat this by the birth–death equations, i.e. $\alpha_n = \alpha,\ n \geqslant 0$; $\beta_1 = \beta$, $\beta_2 = 2\beta$, $\beta_3 = 2\beta$, etc., both servers are busy.

$$p_1 = \frac{\alpha}{\beta} p_0, \qquad p_2 = \frac{\alpha^2}{2\beta^2} p_0, \qquad p_3 = \frac{\alpha^3}{2^2\beta^3} p_0$$

$$p_n = 2\left(\frac{\rho}{2}\right)^n p_0 \qquad \text{where } \rho = \frac{\alpha}{\beta}\ (n = 1, 2, \ldots) \qquad (3.5)$$

Since $\sum_{n=0}^{\infty} p_n = 1$

$$p_0 \left\{ 1 + 2\left[\frac{\rho}{2} + \left(\frac{\rho}{2}\right)^2 + \left(\frac{\rho}{2}\right)^3 + \cdots \right] \right\} = 1$$

$$\therefore \ p_0 \left\{ 1 + \frac{2\dfrac{\rho}{2}}{1 - \dfrac{\rho}{2}} \right\} = 1$$

$$\therefore \ p_0 = \frac{2-\rho}{2-\rho} \tag{3.6}$$

$$p_0 = \frac{2-\rho}{2+\rho}$$

Note for the above $\rho/2 < 1$, i.e. $\rho < 2$. We would intuitively expect this to be the condition for the existence of a steady-state solution.

$$\therefore \ \Pr(N = n) = p_n = 2\left(\frac{2-\rho}{2+\rho}\right)\left(\frac{\rho}{2}\right)^n, \qquad n \geqslant 1$$

$$E[N] = \sum_{n=0}^{\infty} np_n = \sum_{n=1}^{\infty} np_n = 2\left(\frac{2-\rho}{2+\rho}\right)\sum_{n=1}^{\infty} n\left(\frac{\rho}{2}\right)^n$$

$$= 2\left(\frac{2-\rho}{2+\rho}\right)\frac{\dfrac{\rho}{2}}{\left(1 - \dfrac{\rho}{2}\right)^2} = \frac{4\rho}{(2-\rho)^2} \cdot \frac{2-\rho}{2+\rho}$$

$$E[N] = L = \frac{4\rho}{4 - \rho^2} \tag{3.7}$$

The distribution of queueing time

If X denotes the waiting time in the queue with p.d.f. $\varphi(x)$

$$\varphi(x) = \sum_{n=2}^{\infty} 2\left(\frac{2-\rho}{2+\rho}\right)\left(\frac{\rho}{2}\right)^n \frac{(2\beta x)^{n-2} e^{-2\beta x}}{(n-2)!} \, 2\beta, \qquad x > 0$$

A customer will have to queue for time x if on arrival he finds n customers in the system (with $n \geqslant 2$ so that both servers are occupied) and there are $(n-2)$ service completions during time x, when the service rate is 2β (this means our customer is at the head of the queue at time x) and then there is a further service completion in the next instant after x, so that our customer can enter service with one or other of the servers.

We have seen that

$$p_n = 2\left(\frac{2-\rho}{2+\rho}\right)\left(\frac{\rho}{2}\right)^n$$

and the probability for $(n-2)$ service completions in time x is

$$\frac{(2\beta x)^{n-2}e^{-2\beta x}}{(n-2)!}$$

[see equation (0.6)], with $2\beta\delta x$ being the probability for a service completion in $(x, x+\delta x)$. See also the derivation of (2.12).

$$\therefore\ \varphi(x) = 2\left(\frac{2-\rho}{2+\rho}\right)2\beta\left(\frac{\alpha}{2\beta}\right)^2 e^{-2\beta x}\sum_{n=2}^{\infty}\frac{(\alpha x)^{n-2}}{(n-2)!}$$

$$= 4\left(\frac{2-\rho}{2+\rho}\right)\frac{\alpha^2\beta}{4\beta^2}e^{-(2\beta-\alpha)x}$$

$$\therefore\ \varphi(x) = \beta\left(\frac{2-\rho}{2+\rho}\right)\rho^2 e^{-(2\beta-\alpha)x},\qquad x>0$$

$$\Pr(X=0) = p_0 + p_1$$

$$E[X] = \beta\left(\frac{2-\rho}{2+\rho}\right)\frac{\rho^2}{(2\beta-\alpha)^2} = W_q \tag{3.8}$$

The average number in the queue

$$L_q = p_3 + 2p_4 + 3p_5 + \cdots = \frac{2(2-\rho)}{(2+\rho)}\left[\left(\frac{\rho}{2}\right)^3 + 2\left(\frac{\rho}{2}\right)^4 + \cdots\right]$$

$$= 2\left(\frac{2-\rho}{2+\rho}\right)\left(\frac{\rho}{2}\right)^2\left[\frac{\rho}{2} + 2\left(\frac{\rho}{2}\right)^2 + 3\left(\frac{\rho}{2}\right)^3 + \cdots\right]$$

$$= 2\left(\frac{2-\rho}{2+\rho}\right)\left(\frac{\rho}{2}\right)^2\frac{\dfrac{\rho}{2}}{\left(1-\dfrac{\rho}{2}\right)^2} = \frac{\rho^3}{4-\rho^2} = \frac{4\rho}{4-\rho^2} - \rho \tag{3.9}$$

Note $L = L_q + \rho$. Also note that

$$W_q = \beta\left(\frac{2+\rho}{2+\rho}\right)\frac{\rho^2}{(2-\rho)^2\beta^2} = \frac{1}{\beta}\frac{\rho^2}{4-\rho^2} = \frac{1}{\alpha}\frac{\rho^3}{4-\rho^2}$$

$$\therefore\ L_q = \alpha W_q \tag{3.10}$$

We also have that W, the mean time in the system, is given by

$$W = W_q + \frac{1}{\beta} \tag{3.11}$$

$$\therefore \; L = L_q + \rho = \alpha W \qquad (3.12)$$

Equations (3.10) and (3.12) are examples of what is sometimes known as Little's formula. There is no real difficulty in extending these ideas to the $M/M/c$ system; random arrivals at rate α, exponential service at rate β from each of c servers.

We shall have $\alpha_n = \alpha$, $\beta_n = n\beta$ for $n < c$ and $\beta_n = c\beta$ for $n \geq c$. The analysis is straightforward but the algebra becomes complicated.

Example 3.4

Part-finished components from a particular production process are input into one of two identical machines in order to complete their manufacture. The components arrive at random at an average rate of 100/hour and the final processing time on the second machine has a negative exponential distribution with mean 1 minute.

How many components on average are waiting to receive their final processing? Storage problems will arise if the number of components waiting for further processing exceeds ten. How likely is this to happen?

Here we have an $M/M/2$ system, and if we take 1 minute as our time unit, in our usual notation, $\alpha = 100/60 = 5/3$ components per minute and $\beta = 1$. Therefore $\rho = 5/3$ [N.B. $\rho/2 = 5/6 < 1$]. Thus from (3.5) and (3.6)

$$p_0 = \frac{2 - \dfrac{5}{3}}{2 + \dfrac{5}{3}} = \frac{1}{11} \qquad \text{and} \qquad p_n = \frac{2}{11}\left(\frac{5}{6}\right)^n$$

The average number of components in the queue is from (3.9)

$$\left(\frac{5}{3}\right)^3 \bigg/ \left[4 - \left(\frac{5}{3}\right)^2\right] = \frac{125}{33} \simeq 4$$

If M denotes the number in the queue and $\Pr(M = m)$ is denoted by q_m, then

$$q_0 = p_0 + p_1 + p_2 = \frac{1}{11}\left\{1 + \frac{5}{3} + 2\left(\frac{5}{6}\right)^2\right\} = \frac{73}{11 \times 18}$$

$$q_1 = p_3 = \frac{2}{11}\left(\frac{5}{6}\right)^3$$

$$q_2 = p_4 = \frac{2}{11}\left(\frac{5}{6}\right)^4$$

$$q_3 = p_5 = \frac{2}{11}\left(\frac{5}{6}\right)^5 \text{ etc.}$$

Thus

$$\Pr(M \leqslant 10) = q_0 + q_1 + q_2 + \cdots + q_{10}$$

$$= \frac{73}{11 \times 18} + \frac{2}{11} \cdot \frac{125}{216} \left\{ 1 + \left(\frac{5}{6}\right) + \left(\frac{5}{6}\right)^2 + \cdots + \left(\frac{5}{6}\right)^9 \right\}$$

$$= \frac{73}{198} + \frac{2}{11} \cdot \frac{125}{216} \cdot \frac{1 - \left(\dfrac{5}{6}\right)^{10}}{1 - \dfrac{5}{6}}$$

$$= \frac{73}{198} + \frac{125}{198} \left[1 - \left(\frac{5}{6}\right)^{10} \right]$$

$$= 1 - \frac{125}{198} \left(\frac{5}{6}\right)^{10}$$

Thus

$$\Pr(M > 10) = \frac{125}{198} \left(\frac{5}{6}\right)^{10} \approx 0.102$$

Thus the probability is about 10% that this situation will arise. If we put it another way, storage problems will trouble the system for about 10% of the time.

3.5   Examples 7

1   Refer to questions 1 and 2 of Examples 6. If the extra land means that the service station owner can install a second pump and have room for a total of four cars at the station, estimate the profit per week he should make. Does this mean that it is now advisable to rent the land?

2   There are two inspectors who check the work of other workers. The tasks arrive at random at an average rate of 6 pieces/hour and it takes an average of 15 minutes to check each piece with the inspection time having a negative exponential distribution.
    Find the proportion of time that: (i) both inspectors are free; and (ii) a particular inspector is free.
    How long on average must a workman wait until (i) his work is inspected, (ii) he is able to return?

3   A post office counter is divided into two sections, each section being staffed by one clerk. At the first section, which deals with stamps, parcels and pensions, customers arrive at random at an average rate of 18/hour. At the second section, which deals with postal orders, savings and licences, customers arrive at random at an average rate of 12/hour. At each section the service time has a negative exponential distribution with a mean of 2 minutes. At each section estimate the proportion of time the clerk is busy and the average queueing time of an arriving customer.

If the counter is reorganised so that each clerk can deal with the full range of transactions, and customers from a common queue go to the first available clerk, estimate the proportion of time each clerk is busy and the average queueing time of an arriving customer.

3.6   Solutions to Examples 7

1 We now have $a_0 = a_1 = a_2 = a_3 = 1/8$; $a_4 = a_5 = \cdots = 0$;   and $\beta_1 = 1/4$, $\beta_2 = \beta_3 = \beta_4 = 1/2$.

$$p_1 = \tfrac{1}{2} p_0, \; p_2 = \tfrac{1}{8} p_0, \; p_3 = \tfrac{1}{32} p_0, \; p_4 = \tfrac{1}{128} p_0$$

Therefore since $p_0 + p_1 + p_2 + p_3 + p_4 = 1$

$$p_0 = \tfrac{128}{213}, \; p_1 = \tfrac{64}{213}, \; p_2 = \tfrac{16}{213}, \; p_3 = \tfrac{4}{213}, \; p_4 = \tfrac{1}{213}$$

Therefore 212/213 of the potential trade is served.

Average profit per week $= 212/213 \times 1/8 \times 600 \times 7 \times 0.5 - 10 = £251.27$ (>245).
Thus it is sensible to rent the land and install a second pump.

2 α (arrival rate) $= 1/10$ (of a piece/minute) and β (service rate) $= 1/15$ (inspections/minute)

$$\therefore \rho = \frac{\alpha}{\beta} = \frac{3}{2} \; (< 2)$$

$$p_0 = \frac{2 - \rho}{2 + \rho} = \frac{\dfrac{1}{2}}{\dfrac{7}{2}} = \frac{1}{7}$$

$$p_1 = \frac{3}{14}$$

Pr(both inspectors free) $= p_0 = 1/7$.
Pr(a particular inspector is free) $= p_0 + (1/2)p_1 = 1/7 + 3/28 = 1/4$.

$$E[X] = \frac{\dfrac{1}{15} \dfrac{1}{7} \left(\dfrac{3}{2} \right)^2}{\left(\dfrac{2}{15} - \dfrac{1}{10} \right)^2} \quad \text{[see (3.8), formula for } W_q\text{]}$$

$$= \frac{9}{15 \times 7 \times 4} \times 900 = 19.286 \text{ minutes}$$

$$W = W_q + \frac{1}{\beta} = 34.286$$

3 *Before* reorganisation we have two single-server queues. For queue 1, $\alpha = 18/60$ per minute, $\beta = 1/2$, $\rho = 36/60 = 3/5$. Thus using results from (2.8) to (2.15), $p_0 = (1 - \rho) = 2/5$ and proportion of time clerk is busy is $1 - p_0 = 3/5$.

$$W_q = \frac{\alpha}{\beta(\beta - \alpha)} = \frac{3}{5\left(\dfrac{1}{2} - \dfrac{3}{10}\right)} = \frac{3}{5} \cdot \frac{10}{2} = 3 \text{ minutes}$$

For queue 2, $\alpha = 12/60$ per minute, $\beta = 1/2$, $\rho = 2/5$. Therefore, $p_0 = 3/5$ and proportion of time clerk is busy $= 2/5$.

$$W_q = \frac{2}{5\left(\dfrac{1}{2} - \dfrac{2}{10}\right)} = \frac{2}{5} \cdot \frac{10}{3} = \frac{4}{3} \text{ minutes}$$

Note for the two clerks together, on average they are occupied for *half* the total time.

After reorganisation we have a queue with two servers (M/M/2). $\alpha = 30/60 = 1/2$ per minute, $\beta = 1/2$, $\rho = 1$. Thus using results (3.5) to (3.12),

$$p_0 = \frac{2 - \rho}{2 + \rho} = \frac{1}{3}, \qquad p_1 = 2\frac{1}{2}p_0 = \frac{1}{3}$$

Assuming that when there is only one customer in the system, the customer is equally likely to be served by either clerk, the proportion of time that each one of the clerks is idle is

$$p_0 + \tfrac{1}{2}p_1 = \tfrac{1}{3} + \tfrac{1}{6} = \tfrac{1}{2}$$

This is the same on average as before. There is no more work to do.

$$W_q = \frac{1}{\alpha} \frac{\rho^3}{4 - \rho^2} = 2\frac{1}{3} = \frac{2}{3} \text{ minutes}$$

and this is much reduced. Why? The reason is that we have avoided the situation where one clerk is idle whilst the other has a queue of customers waiting for service. This simple example shows why banks, building societies and post offices now adopt this type of system. The work gets spread evenly among the clerks and it reduces waiting times for customers. They also avoid the frustration of 'joining the wrong queue' where the customer in front of them takes an 'age to be served'. Meanwhile others who arrived later but fortuitously joined the 'right queue' have long since been served!

3.7 The machine interference problem

As an illustration, consider an operator who looks after three machines which break down from time to time. When this happens the operator repairs the machine that has stopped. Figure 3.1 illustrates the sort of thing that can happen. The stopped being repaired periods are inevitable but undesirable. The stopped awaiting repair periods are highly undesirable. Furthermore, the machines are only productive during the running periods.

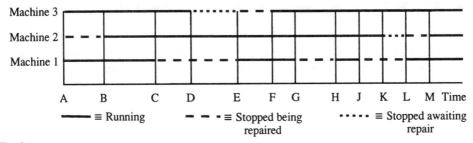

Fig. 3.1

In general if the operator looks after N machines and we consider the system over a long time T hours, during which time the operator works on repairs for T_0 hours (sum of — — — periods) and a total of T_r running hours are achieved (sum of ——— periods)

$$\text{Operative utilisation} = P = T_0/T \qquad (3.13)$$

$$\text{Machine efficiency} = E = T_r/NT \qquad (3.14)$$

The operative utilisation is the proportion of time the operative spends on repairs. The machine efficiency is the proportion of time that a machine is actually running. There is a simple relationship between E and P. If the mean number of stops per unit running time per machine is λ, and the mean repair time is c, then in time T there will be a total of λT_r stops (asymptotically) and it will take a time $\lambda T_r c$ to repair them. Thus

$$T_0 = \lambda T_r c = \rho T_r \qquad \text{where } \rho = \lambda c$$

$$\therefore \ P = \frac{\rho T_r}{T} = N\rho E$$

$$\therefore \ E = \frac{P}{N\rho} \qquad (3.15)$$

If $N = 1$ the operator looks after just one machine. The operative utilisation will be $\rho/(1 + \rho)$ since each hour of running will require ρ hours of repair. The efficiency is just $1/(1 + \rho)$.

If we refer back to (2.2) and (2.3) with $\lambda = \alpha$ and $c = 1/\beta$, then

$$P = p_s = \left(\frac{\alpha}{\alpha + \beta}\right) = \frac{\lambda}{\lambda + \dfrac{1}{c}} = \frac{\lambda c}{1 + \lambda c} = \frac{\rho}{1 + \rho}$$

$$E = p_R = \left(\frac{\beta}{\alpha + \beta}\right) = \frac{\dfrac{1}{c}}{\lambda + \dfrac{1}{c}} = \frac{1}{1 + \lambda c} = \frac{1}{1 + \rho}$$

If N is very large then $P \to 1$. The operator will be kept busy. Then $E \to 1/(N\rho)$ and the average number of machines actually running $\to NE = 1/\rho$.

3.8 The M/M/1 machine interference model (finite source model)

To deal with other values for N we need to make assumptions. If we assume that the machines break down at random at a rate α and that repair times have a negative exponential distribution with mean $1/\beta$, we can model the problem using the birth–death equations.

If $p_n \equiv$ steady-state probability that n machines are stopped (this means that $(N-n)$ are running and could stop), then

$$\alpha_n = \alpha(N-n) \qquad n = 0, 1, 2, ..., N$$
$$\beta_n = \beta \qquad \text{for all values of } n$$

Thus using the general solution (3.3) and (3.4),

$$p_1 = \frac{N\alpha}{\beta} p_0, \qquad p_2 = N(N-1)\left(\frac{\alpha}{\beta}\right)^2 p_0, \qquad p_3 = N(N-1)(N-2)\left(\frac{\alpha}{\beta}\right)^3 p_0 \text{ etc.}$$

$$p_N = N!\left(\frac{\alpha}{\beta}\right)^N p_0$$

$$\therefore p_n = \frac{N!}{(N-n)!}\left(\frac{\alpha}{\beta}\right)^n p_0 \qquad n = 0, 1, 2, ..., N \tag{3.16}$$

To find p_0 we use the condition $\sum_{n=0}^{N} p_n = 1$, which leads to

$$p_0 = 1/\{1 + N\rho + N(N-1)\rho^2 + \cdots + N!\rho^N\} \qquad \text{where } \rho = \frac{\alpha}{\beta} \tag{3.17}$$

It is convenient to denote the denominator above by

$$F(N, \rho) = 1 + N\rho + N(N-1)\rho^2 + N(N-1)(N-2)\rho^3 + \cdots + N!\rho^N$$

We may interpret p_0 as the proportion of time for which all machines are running. The operative utilisation will be $1 - p_0$ and so

$$E = \frac{1-p_0}{N\rho} = \frac{1}{N\rho} \cdot \frac{F(N,\rho)-1}{F(N,\rho)}$$

However, we note that

$$F(N, \rho) = 1 + N\rho[1 + (N-1)\rho + (N-1)(N-2)\rho^2 + \cdots + (N-1)!\rho^{N-1}]$$
$$\therefore F(N, \rho) = 1 + N\rho F(N-1, \rho)$$

We have of course $F(0, \rho) = 1$. Thus

$$E = \frac{F(N-1,\rho)}{F(N,\rho)} \tag{3.18}$$

For a given value of ρ we can use the recurrence to calculate the polynomials $F(M, \rho)$ for $M = 0, 1, 2, ..., N$. Then we can calculate the efficiency as the ratio above. Note that the average number of running machines $= NE$ (see question 1 of Examples 8). The machine interference problem has been solved for other forms of the repair time distribution. The solution for two particular cases can be put in the form below

($\rho = a \times$ mean repair time in both cases). The average number of machines running $= (NY_N)/1 + N\rho YN$ where

$$Y_N = 1 + (N-1)\rho + (N-1)(N-2)\rho^2 + \cdots + (N-1)!\rho^{N-1}$$

for negative exponential repairs, and

$$Y_n = 1 + \binom{N-1}{1}(e^\rho - 1) + \binom{N-1}{2}(e^\rho - 1)(e^{2\rho} - 1) + \cdots$$

$$+ \binom{N-1}{N-1}(e^\rho - 1)(e^{2\rho} - 1)\ldots(e^{(N-1)\rho} - 1)$$

for constant repair time.

The efficiency $E \equiv$ proportion of time a machine is running.

Other quantities of interest are the average time spent stopped (from breakdown to completion of repair) for a machine and the average time a machine has to wait for attention before its repair begins. Let $S \equiv$ average time stopped for a machine. The average time running before a breakdown is $1/a$. [Run times have a negative exponential distribution with mean $1/a$. It is (0.3) again.] Thus

$$E = \frac{\dfrac{1}{a}}{\dfrac{1}{a} + S}$$

whence

$$S = \frac{1}{Ea} - \frac{1}{a}$$

If W is the average time spent waiting for a repair to begin, $S = W + c$ where c is the mean repair time.

$$\therefore W = \frac{1}{Ea} - \frac{1}{a} - c$$

Tables 3.1–3.3 give some numerical results based on some of the formulae just mentioned. The limiting case when N is large and the average number of machines actually running is $1/\rho$ is clearly apparent in the bottom right of Tables 3.1 and 3.3. The corresponding full utilisation of the operator is seen in the bottom right of Table 3.2.

Example 3.5

A weaver has charge of four looms. Yarn breakages occur at random at each loom at an average rate of one every 20 minutes. The time taken to correct a breakage has a negative exponential distribution with mean 1 minute.

Estimate the proportion of time the weaver is actually repairing yarn breakages and the running efficiency of each loom.

We can use (3.16) and (3.17) for this problem with $N = 4$, $a = 1/20$ and $\beta = 1$. Thus

$$p_1 = \frac{4}{20} p_0, \qquad p_2 = \frac{12}{(20)^2} p_0, \qquad p_3 = \frac{24}{(20)^3} p_0, \qquad p_4 = \frac{24}{(20)^4} p_0$$

where $p_n \equiv \Pr(n \text{ looms are stopped})$.

Table 3.1   Average number of machines running exponential repair times

ρ	2	4	N 6	10	15	20
0.01	1.98	3.96	5.94	9.89	14.83	19.76
0.02	1.96	3.92	5.87	9.76	14.60	19.40
0.03	1.94	3.87	5.80	9.61	14.30	18.84
0.04	1.92	3.83	5.72	9.44	13.91	18.00
0.05	1.90	3.78	5.64	9.24	13.40	16.82
0.06	1.88	3.73	5.55	9.01	12.77	15.36
0.08	1.84	3.64	5.36	8.48	11.24	12.33
0.10	1.80	3.53	5.15	7.85	9.64	9.98
0.15	1.71	3.27	4.60	6.23	6.65	6.66
0.20	1.62	3.01	4.04	4.91	4.99	5.00
0.30	1.46	2.52	3.09	3.33	3.33	3.33
0.40	1.32	2.13	2.43	2.50	2.50	2.50

Table 3.2   Operative utilisation (%)

ρ	2	4	N 8	10	14
0.10	18.0	35.3	66.2	78.5	94.3
0.12	21.2	41.2	74.6	86.3	97.8
0.14	24.2	46.5	81.3	91.6	99.2
0.16	27.1	51.5	86.4	94.9	99.7
0.18	29.8	56.0	90.2	97.3	99.9
0.20	32.4	60.2	93.0	98.2	100
0.40	52.8	85.0	99.7	100	100

Table 3.3   Average number of machines running constant repair times

ρ	2	4	N 6	10	15	20
0.01	1.98	3.96	5.94	9.90	14.84	19.78
0.02	1.96	3.92	5.88	9.78	14.65	19.50
0.03	1.94	3.88	5.81	9.66	14.42	19.10
0.04	1.92	3.84	5.74	9.52	14.13	18.48
0.05	1.90	3.79	5.67	9.37	13.75	17.45
0.06	1.88	3.75	5.60	9.19	13.23	15.93
0.08	1.84	3.67	5.45	8.76	11.72	12.48
0.10	1.81	3.58	5.28	8.21	9.89	9.99
0.15	1.72	3.35	4.80	6.48	6.66	6.66
0.20	1.64	3.12	4.26	4.99	5.00	5.00
0.30	1.49	2.67	3.23	3.33	3.33	3.33
0.40	1.36	2.25	2.48	2.50	2.50	2.50

To find p_0 we have

$$p_0\left[1 + \frac{4}{20} + \frac{12}{400} + \frac{24}{8000} + \frac{24}{160\,000}\right] = 1$$

$$\therefore \; p_0 = \frac{1}{1.23315} \approx 0.811$$

Thus operative utilisation $\approx 1 - 0.811 = 0.189$.
The average number of looms running is given by

$$4p_0 + 3p_1 + 2p_2 + p_3 = 3.78$$

This of course accords with the value given in Table 3.1 corresponding to $\rho = 0.05$ and $N = 4$. The efficiency for each loom is $3.78/4 = 94.5\%$ and is the long run proportion of time each one is actually producing material.

3.9   Examples 8

1   One operator looks after N identical machines. Each machine is independent of the others and breaks down at random in running time at an average rate α. Repair times have an exponential distribution with mean $1/\beta$. Obtain, in terms of ρ, an expression for p_n, the probability that n machines are stopped. Find an expression for the average number of machines running.

2   For the situation above show that the proportion of time that a machine spends waiting for repair is

$$\frac{(N-1)\rho^2 + 2(N-1)(N-2)\rho^3 + 3(N-1)(N-2)(N-3)\rho^4 + \cdots + (N-1)(N-1)!\rho^N}{1 + N\rho + N(N-1)\rho^2 + \cdots + N!\rho^N}$$

[Hint: a machine is in one of three possible states: (i) running, (ii) being repaired, (iii) waiting to be repaired.]

3   In the situation above write down in terms of N and α the average length of an operative's idle period. In a long time T, what proportion of the time is the operative (i) idle, (ii) busy? Find how many idle periods there are in time T. How many busy periods are there in time T? Find an expression for the average length of a busy period.

Solutions to examples 8

1   From (3.16) and (3.17),

$$p_n = \frac{N(N-1)\dots(N-n+1)\rho^n}{1 + N\rho + N(N-1)\rho^2 + \cdots + N!\rho^N}$$

Average number running $= \sum_{n=0}^{N} (N-n)p_n$

$$= \frac{N.1 + (N-1)N\rho + (N-2)N(N-1)\rho^2 + \cdots + (N-N)N!\rho^N}{1 + N\rho + N(N-1)\rho^2 + \cdots + N!\rho^N}$$

$$= \frac{N[1 + (N-1)\rho + (N-1)(N-2)\rho^2 + \cdots + (N-1)!\rho^{N-1}]}{1 + N\rho + N(N-1)\rho^2 + \cdots + N!\rho^N}$$

$$= N\frac{F(N-1,\rho)}{F(N,\rho)} = NE$$

with E as given by (3.18).

2 Proportion of time running $= (1-p_0)/N\rho \ (\equiv E)$
Proportion of time being repaired $= (1-p_0)/N \ (\equiv P/N)$
Proportion of time spent waiting to be repaired is

$$1 - \frac{(1-p_0)}{N} - \frac{(1-p_0)}{N\rho}$$

$$= 1 - \frac{(1-p_0)(1+\rho)}{N\rho}$$

$$= 1 - \frac{\{1 + (N-1)\rho + (N-1)(N-2)\rho^2 + \cdots + (N-1)!\rho^{N-1}\}(1+\rho)}{1 + N\rho + N(N-1)\rho^2 + \cdots + N!\rho^N}$$

$$= \frac{(N-1)\rho^2 + 2(N-1)(N-2)\rho^3 + 3(N-1)(N-2)(N-3)\rho^4 + \cdots + (N-1)(N-1)!\rho^N}{1 + N\rho + N(N-1)\rho^2 + \cdots + N!\rho^N}$$

after some algebra.

Alternative solution
Consider a particular machine. Given that n machines are stopped, the probability that this machine is one of them is n/N and the probability that it is one of the $(n-1)$ machines out of the n that are waiting is $(n-1)/n$.

$\therefore$ Pr(this machine is stopped waiting $= \sum_{n=2}^{N} p_n \frac{n}{N} \cdot \frac{n-1}{n}$

$$= \frac{1}{N} \sum_{n=2}^{N} (n-1)p_n$$

$$= \frac{N(N-1)\rho^2 + N(N-1)(N-2)2\rho^3 + N(N-1)(N-2)(N-3)3\rho^4 + \cdots + N!(N-1)\rho^N}{N\{1 + N\rho + N(N-1)\rho^2 + \cdots + N!\rho^N\}}$$

$$= \frac{(N-1)\rho^2 + 2(N-1)(N-2)\rho^3 + 3(N-1)(N-2)(N-3)\rho^4 + \cdots + (N-1)(N-1)!\rho^N}{1 + N\rho + N(N-1)\rho^2 + \cdots + N!\rho^N}$$

3 An operator's idle period begins when all N machines are running and ends as soon as one stops. Since the breakdown rate when all N machines are running is $N\alpha$, the time until the first breakdown has a negative exponential distribution with mean $1/N\alpha$.

In a long time T the operator is idle for time $p_0 T = T/F(N, \rho)$.

$$\text{The operator is busy for time } (1 - p_0)T = \frac{[F(N, \rho) - 1]T}{F(N, \rho)}$$

Thus the number of idle periods will be (on average) $p_0 T/(1/N\alpha) = N\alpha p_0 T$ and this will equal the number of busy periods, since idle and busy periods alternate. Then the average length of a busy period is

$$\frac{(1 - p_0)T}{N\alpha p_0 T} = \frac{F(N, \rho) - 1}{N\alpha} = \frac{(F(N, \rho) - 1)}{N\rho\beta}$$

$$= \frac{1}{\beta} F(N - 1, \rho)$$

4
The M/G/1 system

We have a queue with one server, random arrivals (at average rate α) and general independent service times. Let the service time have probability density function $b(t)$; $t \geq 0$. It is only in the case of exponential service that the probability that a service (which is in progress) will terminate in an interval of length δt, is independent of the time for which the service has been in progress. The result (0.5) is unique to negative exponential service times with mean $1/\beta$. For other distributions this is not true and we must develop other approaches, which we do in this and the next chapter.

Label the customers in sequence $1, 2, 3, \ldots, n, \ldots$. Let q_n denote the number of customers in the system at the moment that the service of customer n finishes. Suppose ζ_n customers arrive during the service time of customer n. q_n and ζ_n are random variables and $\zeta_1, \zeta_2, \zeta_3, \ldots, \zeta_n$ are independent and identically distributed random variables. Furthermore, ζ_n is independent of $q_1, q_2, \ldots, q_{n-1}$.

$$q_{n+1} = q_{n-1} + \zeta_{n+1} \qquad (q_n > 0)$$
$$= \zeta_{n+1} \qquad (q_n = 0)$$

The case where $q_n > 0$ is illustrated in Fig. 4.1.

Customer $(n+1)$ leaves behind the customers left behind by customer n less customer $(n+1)$ himself as well as those who arrive during the service time of customer $(n+1)$. When $q_n = 0$, customer $(n+1)$ finds an empty system and so goes straight into service. Customer $(n+1)$ leaves behind only those who arrive during the service time of customer $(n+1)$ in this case.

$$q_{n+1} = q_n - U(q_n) + \zeta_{n+1} \qquad \text{where } U(q_n) = 1; \quad q_n > 0$$
$$= 0; \quad q_n = 0 \qquad (4.1)$$

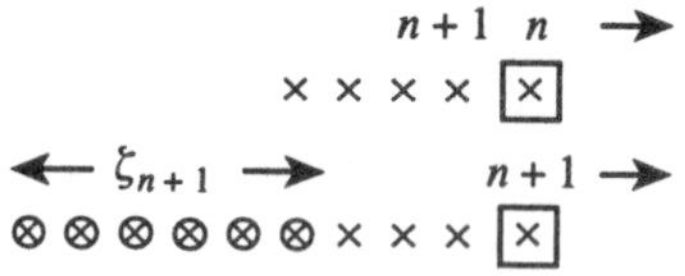

$$
\begin{array}{cc}
n+1 \quad n & \rightarrow \\
\times \times \times \times \boxed{\times} &
\end{array}
$$

$$
\begin{array}{cc}
\leftarrow \zeta_{n+1} \rightarrow & n+1 \rightarrow \\
\otimes \otimes \otimes \otimes \otimes \times \times \times \boxed{\times} &
\end{array}
$$

Fig. 4.1

Now if the service time of customer n is s, ζ_n is a Poisson variable with parameter αs

$$\Pr(\zeta_n = k \mid s) = \frac{e^{-\alpha s}(\alpha s)^k}{k!} \qquad (4.2)$$

$$\therefore \ E[\zeta_n \mid s] = \alpha s, \qquad E[\zeta_n^2 \mid s] = \alpha s + (\alpha s)^2,$$

$$\therefore \ E[\zeta_n] = \int_0^\infty \alpha s b(s) \, ds = \alpha c = \rho$$

where c is the mean service time

$$E[\zeta_n^2] = ac + a^2 E[S^2] = \rho + a^2\{c^2 + \text{Var}[S]\}$$
$$\therefore \quad E[\zeta_n^2] = \rho + \rho^2 + a^2 \text{Var}[S]$$

Now

$$E[q_{n+1}] = E[q_n] - E[U(q_n)] + E[\zeta_{n+1}]$$

If steady-state conditions exist as $n \to \infty$ so that $\text{Limit}_{n \to \infty} \Pr(q_n = k) = \pi_k$

$$E[q_n] = E[q_{n+1}]$$
$$\therefore \quad E[U(q_n)] = E[\zeta_{n+1}] = \rho \tag{4.3}$$

Note $E[U(q_n)] = \Pr(q_n \neq 0) = \rho$

$$\therefore \quad \Pr(q_n = 0) = 1 - \rho = \pi_0 = \Pr(\text{system is left empty by a customer who leaves}) \tag{4.4}$$

Note we must have $\rho < 1$ for a steady state to exist. On squaring the fundamental equation (4.1),

$$q_{n+1}^2 = q_n^2 + \{U(q_n)\}^2 + \zeta_{n+1}^2 + 2q_n\zeta_{n+1} - 2\zeta_{n+1}U(q_n) - 2q_nU(q_n)$$
$$= q_n^2 + U(q_n) + \zeta_{n+1}^2 + 2q_n\zeta_{n+1} - 2\zeta_{n+1}U(q_n) - 2q_n$$

since $q_nU(q_n) = q_n$ and $\{U(q_n)\}^2 = U(q_n)$

$$\therefore \quad E[q_{n+1}^2] = E[q_n^2] + E[U(q_n)] + E[\zeta_{n+1}^2] + 2E[q_n\zeta_{n+1}] - 2E[\zeta_{n+1}U(q_n)] - 2E[q_n]$$

But since q_n and ζ_{n+1} are independent

$$E[q_n\zeta_{n+1}] = E[q_n]E[\zeta_{n+1}]$$
$$E[\zeta_{n+1}U(q_n)] = E[\zeta_{n+1}]E[U(q_n)]$$

Now in the steady-state situation $E[q_{n+1}^2] = E[q_n^2]$

$$\therefore \quad 0 = \rho + \rho + \rho^2 + a^2 \text{Var}[S] + 2\rho E[q_n] - 2\rho^2 - 2E[q_n]$$

$$\therefore \quad E[q_n] = \frac{\rho + \rho + \rho^2 + a^2 \text{Var}[S] - 2\rho^2}{2(1 - \rho)}$$

$$\therefore \quad E[q_n] = \rho + \frac{\rho^2 + a^2 \text{Var}[S]}{2(1 - \rho)} \tag{4.5}$$

Thus the average value of q_n increases with the mean service time ($\rho = ac$) and with the variance of the service time.

We consider the M/M/1 system with $b(t) = 1/c(e^{-t/c})$. This type of service has mean c and variance c^2.

$$\therefore \quad E[q_n] = \rho + \frac{2\rho^2}{2(1 - \rho)} = \frac{\rho}{1 - \rho} = L \qquad \text{(as given at (2.10))}$$

In the case of constant service time (M/D/1) with mean c and $\text{Var}[S] = 0$

$$E[q_n] = \rho + \frac{\rho^2}{2(1 - \rho)} = \frac{\rho}{1 - \rho}\left[1 - \frac{1}{2}\rho\right]$$

Thus for M/M/1 analysed in this way $E[q_n] = L$.

The mean number in the system at the moment when a customer leaves the system = the mean number in the system at an arbitrary moment. If the nth customer waits in the queue for time X

$$E[q_n | X \text{ and } S] = a(X + S) = aY$$

$$\therefore \ E[q_n] = aE[X] + aE[S]$$

$$\therefore \ \rho + \frac{\rho^2 + a^2 \, \text{Var}[S]}{2(1 - \rho)} = aW_q + \rho$$

$$\therefore \ aW_q = \frac{\rho^2 + a^2 \, \text{Var}[S]}{2(1 - \rho)} = L_q \tag{4.6}$$

where W_q is the mean queueing time and L_q the mean number in the queue. Similarly,

$$E[q_n] = aE[Y] \tag{4.7}$$

i.e. $L = aW$ where W is the mean time spent in the system.

Compare these results with (2.11) and (2.15). Results (4.6) and (4.7) are particular cases of Little's law, i.e. the average number in a system is equal to the average arrival rate to the system multiplied by the average time spent in the system. With regard to (4.5) and (4.6) we observe that both the number and waiting time of customers increases with ρ and the variance of the service time. Thus we can reduce these by reducing both the mean service time (i.e. ρ) and its variance. Indeed in the comparison between the M/M/1 system and the M/D/1 system [deterministic, i.e. constant service time] following (4.5) we see that in the second case the average number of customers has been reduced by a factor $[1 - \rho/2]$. If ρ is small this is only just under 1, but as ρ gets near to 1 this factor decreases to 1/2. We can almost halve the number waiting and their waiting time by cutting out all variability in the service time in this case. Of course we still have to implement the theory in practice to actually do this.

Example 4.1

A large factory has a medical centre to deal with minor injuries. Records kept over several years show that, when the centre is open, patients arrive randomly at an average rate of four per hour. The time taken by the nurse at the centre to deal with a patient has a distribution with mean 10 minutes and variance 10 minutes. Find the proportion of time that the nurse is free and calculate the mean number of patients at the centre (i.e. waiting or being treated).

If a clerk were employed at the centre to take care of the administrative details, the treatment time would be reduced but the cost of the clerk would be £250 per week. It is estimated that the mean treatment time would be reduced to 6 minutes with no change in the variance. The cost of a worker being temporarily absent from his post is £8/hour when a 40-hour week is worked. Does it seem financially viable to employ the clerk?

With one minute as our time unit we have an M/G/1 system with (in the notation of this section) $a = 1/15$, $c = 10$ (minutes), so that $\rho = 2/3$. We only need the mean and variance of the service time (not the full distribution) in order to calculate the mean number left in the system. On using (4.4)

$$\Pr(\text{nurse is left free}) = \Pr(q_n = 0) = 1 - \rho = \tfrac{1}{3}$$

and this can be interpreted as the proportion of time the nurse is free. By (4.5)

$$E[q_n] = \frac{2}{3} + \frac{\dfrac{4}{9} + \dfrac{10}{225}}{2\left(1 - \dfrac{2}{3}\right)}$$

$$= \frac{2}{3} + \frac{2}{3} + \frac{30}{450} = \frac{21}{15} = \frac{7}{5}$$

With this arrangement the cost of keeping patients waiting and away from their posts is $7/5 \times 40 \times 8 = £448$ each week.

If the reduction in treatment time is as anticipated, we shall have a second M/G/1 system with $\alpha = 1/15$, $c' = 6$ and $\rho' = 2/5$.

$$E[q'_n] = \frac{2}{5} + \frac{\dfrac{4}{25} + \dfrac{10}{225}}{2 \times \left(1 - \dfrac{2}{5}\right)}$$

$$= \frac{2}{5} + \frac{2}{15} + \frac{1}{27} = \frac{54 + 18 + 5}{5 \times 27} = \frac{77}{135}$$

Thus the total costs associated with the new arrangement are $250 + 8 \times 40 \times 77/135 = £432.52$.

Thus the new arrangement is financially viable and will reduce the number of workers away from their work.

4.1 Examples 9

1 For the M/G/1 queue if

$$b(t) = \frac{k\beta(k\beta t)^{k-1} e^{-k\beta t}}{(k-1)!}; \qquad t \geq 0$$

show that mean service time $= 1/\beta$; variance of service time $= 1/(k\beta^2)$. Use the results (4.5), (4.6) to show that

(i) $L = \dfrac{\alpha}{\beta} + \dfrac{k+1}{2k} \dfrac{\alpha^2}{\beta(\beta - \alpha)}$

(ii) $L_q = \dfrac{k+1}{2k} \dfrac{\alpha^2}{\beta(\beta - \alpha)}$

(iii) $W = \dfrac{k+1}{2k} \dfrac{\alpha}{\beta(\beta - \alpha)} + \dfrac{1}{\beta}$

(iv) $W_q = \dfrac{k+1}{2k} \dfrac{\alpha}{\beta(\beta - \alpha)}$

2 By letting $k \to \infty$ in question 1, find corresponding results for the case of constant service time with value $1/\beta$.

3 An aeroplane takes almost exactly 5 minutes to land after it has been given the signal to land by traffic control. Aeroplanes arrive at random at an average rate of 6/hour. How long can a plane expect to circle before getting the signal to land?

4 Orders from a warehouse are sent to customers by lorry. The orders arrive at random at a rate of 0.9/day. The time to load and deliver to the client has p.d.f. $b(t) = 4te^{-2t}$. What is the mean time between the receipt of an order and delivery to a customer?

4.2   Solutions to examples 9

1
$$E[T] = \int_0^\infty \frac{tk\beta(k\beta t)^{k-1}e^{-k\beta t}}{(k-1)!}\, dt = \frac{1}{k\beta}\int_0^\infty \frac{y^k e^{-y}}{(k-1)!}\, dy \qquad [\text{put } y = k\beta t]$$

$$= \frac{1}{k\beta}\frac{k!}{(k-1)!} = \frac{1}{\beta}$$

$$E[T^2] = \frac{1}{(k\beta)^2}\int_0^\infty \frac{y^{k+1}e^{-y}}{(k-1)!}\, dy = \frac{k+1}{k\beta^2} = \frac{1}{\beta^2} + \frac{1}{k\beta^2}$$

$$\therefore \; \text{Var}[T] = \frac{1}{k\beta^2}$$

Thus $\rho = \alpha \times E[T] = \alpha/\beta$

$$\therefore \; L = \frac{\alpha}{\beta} + \frac{\dfrac{\alpha^2}{\beta^2} + \dfrac{\alpha^2}{k\beta^2}}{2\left(1 - \dfrac{\alpha}{\beta}\right)} = \frac{\alpha}{\beta} + \frac{k+1}{2k}\frac{\alpha^2}{\beta(\beta - \alpha)}$$

$$L_q = L - \rho = \frac{k+1}{2k}\frac{\alpha^2}{\beta(\beta - \alpha)}$$

$$W_q = \frac{L_q}{\alpha}, \qquad W = \frac{L}{\alpha}$$

Thus

$$W_q = \frac{k+1}{2k}\frac{\alpha}{\beta(\beta - \alpha)}, \qquad W = \frac{1}{\beta} + \frac{k+1}{2k}\frac{\alpha}{\beta(\beta - \alpha)}$$

2 As $k \to \infty$, $E[T] = 1/\beta$ and $\text{Var}[T] = 1/(k\beta^2) \to 0$. Thus this corresponds to constant service (mean) $1/\beta$. As $k \to \infty$, $k + l/(2k) \to 1/2$

$$\therefore L = \frac{\alpha}{\beta} + \frac{\alpha^2}{2\beta(\beta - \alpha)}$$

$$L_q = \frac{\alpha^2}{2\beta(\beta - \alpha)}$$

$$W = \frac{1}{\beta} + \frac{\alpha}{2\beta(\beta - \alpha)}$$

$$W_q = \frac{\alpha}{2\beta(\beta - \alpha)}$$

3 $\alpha = 1/10$ arrivals/minute, $E[S] = c = 5$ and $\text{Var}[S] = 0$.

$$\therefore \rho = \alpha c = \frac{1}{2} \qquad (< 1)$$

$$\therefore \alpha W_q = \frac{\rho^2 + \alpha^2 \, \text{Var}[S]}{2(1 - \rho)} \qquad ((4.6))$$

$$\therefore \frac{1}{10} W_q = \frac{\dfrac{1}{4}}{2 \cdot \dfrac{1}{2}} = \frac{1}{4}$$

$$\therefore \ W_q = 2\tfrac{1}{2} \text{ minutes}$$

4 $b(t) = 4te^{-2t}; \qquad t \geq 0$

$$E[T] = \int_0^\infty 4t^2 e^{-2t} \, dt = \int_0^\infty \frac{z^2 e^{-z}}{2} \, dz = 1 \qquad [z = 2t]$$

$$E[T^2] = \int_0^\infty 4t^3 e^{-2t} \, dt = \int_0^\infty \frac{z^3 e^{-z}}{4} \, dz = \frac{3}{2}$$

$$\therefore \ \text{Var}[T] = \text{Var}[S] = \frac{1}{2}$$

$$\therefore \ \alpha = 0.9, \qquad c = 1, \qquad \rho = 0.9$$

$$\therefore \ \alpha W = \rho + \frac{\rho^2 + \alpha^2 \, \text{Var}[S]}{2(1 - \rho)} \qquad ((4.5) \text{ and } (4.7))$$

$$\therefore \ 0.9W = 0.9 + \frac{0.81 + 0.81\left(\frac{1}{2}\right)}{2(1 - 0.9)} = 0.9 + \frac{1.215}{0.2} = 0.9 + 6.075$$

$$\therefore \ 0.9W = 6.975$$

$$\therefore \ \ W = 7.75 \text{ days.}$$

4.3 The steady-state distribution of q_n

Limit$_{n \to \infty}$ $\Pr(q_n = k) = \pi_k$ is used to denote the steady-state probabilities for the distribution of q_n. We define the probability generating function (p.g.f.)

$$Q(z) = \sum_{k=0}^{\infty} \pi_k z^k = E[z^{q_n}] = E[z^{q_{n+1}}]$$

$$\therefore \ Q(z) = E\{z^{q_n - U(q_n) + \zeta_{n+1}}\}$$

$$= E\{z^{q_n - U(q_n)}\} E\{z^{\zeta_{n+1}}\}$$

since q_n and ζ_{n+1} are independent.

$$\therefore \ \ Q(z) = E\{z^{q_n - U(q_n)}\} \Phi(z)$$

where $\Phi(z)$ is the p.g.f. of ζ_{n+1}. Now

$$E\{z^{q_n - U(q_n)}\} = \pi_0 + \pi_1 z^{1-1} + \pi_2 z^{2-1} + \pi_3 z^{3-1} + \cdots$$

$$= \pi_0 + \sum_{r=1}^{\infty} \pi_r z^{r-1}$$

$$= (1 - \rho) + \frac{Q(z) - (1 - \rho)}{z}$$

since $\pi_0 = 1 - \rho$ (from (4.4)).

$$\therefore \ Q(z) = \Phi(z)\left[(1 - \rho) + \frac{Q(z) - (1 - \rho)}{z}\right]$$

Thus $zQ(z) = z\Phi(z)(1 - \rho) + \Phi(z)Q(z) - (1 - \rho)\Phi(z)$

$$\therefore \ Q(z) = \frac{(1 - \rho)(1 - z)\Phi(z)}{\Phi(z) - z} \qquad (4.8)$$

Now

$$\Phi(z) = E\{z^{\zeta_n}\} = \sum_{k=0}^{\infty} z^k \Pr(\zeta_n = k)$$

and

$$\Pr(\zeta_n = k) = \int_0^\infty \frac{e^{-ax}(ax)^k}{k!} b(x) \, dx \qquad \text{(from (4.2))}$$

$$\therefore \Phi(z) = \sum_{k=0}^\infty z^k \int_0^\infty \frac{e^{-ax}(ax)^k}{k!} b(x) \, dx$$

$$= \int_0^\infty e^{-ax} \left(\sum_{k=0}^\infty \frac{(axz)^k}{k!} \right) b(x) \, dx$$

$$\therefore \Phi(z) = \int_0^\infty e^{-ax(1-z)} b(x) \, dx$$

$$= B^*[\alpha(1-z)] \qquad (4.9)$$

where

$$B^*(s) = \int_0^\infty e^{-st} b(t) \, dt \text{ is the Laplace transform of } b(t)$$

Thus, provided we can find the Laplace transform of the probability density function for the service time (in principle just a matter of integration or table look-up), we can find the probability generating function $\Phi(z)$ by using (4.9). Then using (4.8) we can find $Q(z)$. The power series expansion of this will give π_n as the coefficient of z^n. Further $Q'(1)$, $Q''(1)$, etc. will allow us to evaluate the mean and variance of the number in the system at the moments of service completion.

In the case of the M/M/1 system things are fairly easy.
$b(t) = \beta e^{-\beta t}$ for negative exponential service at rate β.

$$B^*(s) = \int_0^\infty \beta e^{-\beta t} e^{-st} \, dt = \frac{\beta}{\beta + s}$$

$$\therefore \Phi(z) = \frac{\beta}{\beta + \alpha(1-z)} = \frac{1}{1 + \rho(1-z)} \qquad \text{where } \rho = \frac{\alpha}{\beta}$$

$$\therefore Q(z) = \frac{(1-\rho)(1-z)}{1 + \rho(1-z)} \cdot \frac{1}{\dfrac{1}{1 + \rho(1-z)} - z}$$

$$\therefore Q(z) = \frac{(1-\rho)(1-z)}{1 + \rho(1-z)} \cdot \frac{1 + \rho(1-z)}{1 - z - \rho z(1-z)} = \frac{(1-\rho)}{1 - \rho z}$$

$$\therefore Q(z) = (1-\rho)[1 + \rho z + \rho^2 z^2 + \cdots]$$

$$\therefore \pi_n = (1-\rho)\rho^n \qquad [= p_n \text{ as before, (2.6) and (2.7)}]$$

The number in the system at the moments when a service finishes has the same distribution as the number in the system at an arbitrary moment.

4.4   The time spent in the system (steady state)

Let the time the customers spend in the system (queueing time plus service time; $X + S$) be a random variable Y with p.d.f. $f(y)$. Then q_n is the number of customers who arrive during the time in the system of customer n.

$$\Pr(q_n = k) = \pi_k = \int_0^\infty \frac{e^{-ay}(ay)^k}{k!} f(y)\, dy$$

$$\therefore\ Q(z) = \sum_{k=0}^\infty z^k \int_0^\infty \frac{e^{-ay}(ay)^k}{k!} f(y)\, dy$$

$$= \int_0^\infty e^{-ay} f(y) \left(\sum_{k=0}^\infty \frac{(ayz)^k}{k!} \right) dy$$

$$= \int_0^\infty e^{-ay(1-z)} f(y)\, dy$$

$$\therefore\ Q(z) = F^*(a(1-z)) \qquad \text{where } F^*(s) = \int_0^\infty e^{-sy} f(y)\, dy \qquad (4.10)$$

There is the same relationship between $Q(z)$ and $f(y)$ as exists between $\Phi(z)$ and $b(t)$. Compare (4.10) with (4.9). But

$$Q(z) = \frac{(1-\rho)(1-z)B^*(a(1-z))}{B^*(a(1-z)) - z} = F^*(a(1-z))$$

Thus if we replace $a(1-z)$ by s, i.e. $z = 1 - s/a$

$$F^*(s) = \frac{(1-\rho)\dfrac{s}{a}B^*(s)}{B^*(s) + \dfrac{s}{a} - 1}$$

$$\therefore\ F^*(s) = \frac{(1-\rho)sB^*(s)}{s - a + aB^*(s)} \qquad (4.11)$$

and in principle if we know $b(t)$ we can find $B^*(s)$ and hence $F*(s)$ and also $f(y)$.

Again in the case of negative exponential service $b(t) = \beta e^{-\beta t}$ we can get exact results. $B^*(s) = \beta/(\beta + s)$.

$$\therefore\ F^*(s) = \frac{(\beta - a)s\beta}{\dfrac{\beta(\beta + s)}{s - a + \dfrac{a\beta}{\beta + s}}} = \frac{\beta - a}{\beta - a + s}$$

$$\therefore\ f(y) = (\beta - a)e^{-(\beta - a)y}; \qquad y \geq 0 \qquad \text{[see Examples 5, question 4]}$$

If X represents the time spent in the queue, then $Y = X + S$, where S is the service time and is independent of X.

$$\therefore\ F^*(s) = V^*(s)B^*(s)$$

where $V^*(s)$ is the Laplace transform of the queueing time distribution

$$\therefore \ V^*(s)B^*(s) = F^*(s) = \frac{(1-\rho)sB^*(s)}{s - \alpha + \alpha B^*(s)}$$

$$\therefore \ V^*(s) = \frac{(1-\rho)s}{s - \alpha + \alpha B^*(s)} \tag{4.12}$$

Care is needed here since the distribution of X is part discrete and part continuous.

$$\Pr(X = 0) = \pi_0 = (1-\rho),$$

otherwise X has p.d.f. $\varphi(x)$; $x > 0$.

$$V^*(s) = \Pr(X = 0)e^{-s0} + \int_0^\infty e^{-sx}\varphi(x) \, dx$$

$$= (1-\rho) + \int_0^\infty e^{-sx}\varphi(x) \, dx$$

Of course we only get Laplace transforms. There remains an inversion problem. Note if

$$F^*(s) = \int_0^\infty f(y)e^{-sy} \, dy \text{ then } f(y) = \frac{1}{2\pi i} \int_{c-i\infty}^{c+i\infty} e^{sy}F^*(s) \, ds$$

where the line from $c - i\infty$ to $c + i\infty$ is to the right of all singularities of $F^*(s)$. If we cannot do the inversion we can still find the moments of Y since

$$F^*(s) = 1 - \mu_1 s + \frac{\mu_2 s^2}{2!} - \frac{\mu_3 s^3}{3!} + \cdots$$

where

$$\mu_r = \int_0^\infty y^r f(y) \, dy = \mathrm{E}[Y^r]$$

Example 4.2

Consider the car-wash in Example 2.3. Here cars arrived at random at an average rate of 5/hour. Suppose that instead of self-service the wash is automated so that each wash takes exactly 6 minutes.

Find the generating function for the number of cars left at the wash by a departing customer and hence find the mean number of cars left. Find also the Laplace transform of the time spent by a car waiting to be washed and find the mean value of this time.

In this case $\alpha = 1/12$ of a car/minute and the service time is constant at 6 minutes with probability density function $b(t) = \delta(t - 6)$, where $\delta(x)$ is the Dirac delta function.

$$\therefore \ B^*(s) = \int e^{-st} \, \delta(t - 6) \, dt = e^{-6s}$$

Thus using the result (4.9) and the notation of this chapter

$$\Phi(z) = e^{-6(1-z)/12} = e^{-(1-z)/2}$$

$\rho = 1/2$, so that from (4.8) the required probability generating function is

$$Q(z) = \frac{\dfrac{1}{2}(1-z)e^{-(1-z)/2}}{e^{-(1-z)/2} - z}$$

To find π_n we would need to expand $Q(z)$ as a power series in z when π_n would be the coefficient of z^n. The algebra to do this would be messy. The mean value is given by $Q'(1)$, and it is clear that if we simply differentiate $Q(z)$ and then substitute $z = 1$ we will obtain an indeterminate form $0/0$ and l'Hopital's rule will be needed.

An alternative (see Chapter 1) is to expand $Q(z)$ as a power series in y where $y = z - 1$. The mean is the coefficient of y in this expansion. In this case

$$Q(z) = R(y) = \frac{\dfrac{1}{2} y e^{y/2}}{1 + y - e^{y/2}}$$

$$R(y) = \frac{\dfrac{1}{2} y\left(1 + \dfrac{y}{2} + \dfrac{y^2}{8} + \cdots\right)}{1 + y - 1 - \dfrac{y}{2} - \dfrac{y^2}{8} \cdots}$$

$$= \frac{\dfrac{1}{2} y\left(1 + \dfrac{y}{2} + \dfrac{y^2}{8}\right)}{\dfrac{y}{2} - \dfrac{y^2}{8} - \cdots}$$

$$= \left(1 + \dfrac{y}{2} + \dfrac{y^2}{8} \cdots\right)\left(1 - \dfrac{y}{4} \cdots\right)^{-1}$$

$$= \left(1 + \dfrac{y}{2} + \dfrac{y^2}{8} + \cdots\right)\left(1 + \dfrac{y}{4} \cdots\right)$$

$$= 1 + \dfrac{3y}{4} + \cdots$$

to first degree in y. The algebra here is quite messy. Thus the mean is $3/4$. In this case it might be simpler to use (4.5), since $\rho = 1/2$ and the service time has zero variance. Thus

$$E[q_n] = \frac{1}{2} + \frac{\dfrac{1}{4}}{2\left(1 - \dfrac{1}{2}\right)} = \frac{1}{2} + \frac{1}{4} = \frac{3}{4}$$

as before.

For the Laplace transform of the time in the queue we have from (4.12)

$$V^*(s) = \frac{\dfrac{1}{2} s}{s - \dfrac{1}{12} + \dfrac{1}{12} e^{-6s}} = \frac{6s}{12s - 1 + e^{-6s}}$$

Again the inversion of this will present serious problems. However, the mean time in the queue can be determined from the power series expansion of $V^*(s)$. We need the negative of the coefficient of s in this expansion.

$$V^*(s) = \frac{6s}{12s - 1 + 1 - 6s + \dfrac{36s^2}{2} + \cdots}$$

$$= \frac{6s}{6s(1 + 3s + \cdots)}$$

$$= (1 + 3s + \cdots)^{-1}$$

$$= 1 - 3s + \cdots$$

Thus the mean time in the queue is 3 minutes.

By considering the expansion of $V^*(s)$ to higher powers of s we could find the variance of the queueing time. Of course if we just want the mean it is probably simpler to proceed directly from (4.6):

$$\alpha W_q = \frac{\rho^2}{2(1 - \rho)}$$

where $\alpha = 1/12$ and $\rho = 1/2$. Thus $W_q = 3$ minutes, as before.

It should be noted that the automatic car-wash is much more efficient than the customer self-service system and this is because the variance of the service time has been reduced to zero. The average number in the system has been reduced from 1 to 3/4; the average queueing time from 6 minutes to 3 minutes (see Example 2.3).

4.5 Examples 10

1 Consider a single-server queue with random arrivals at rate α and constant service time $1/\beta$ (to compare with negative exponential service at rate β). Show that (in the notation of this chapter)

$$B^*(s) = e^{-(s/\beta)}$$

Find an expression for $V^*(s)$ and deduce the mean and variance of the queueing time.

2 Use the result (4.12) to find $V^*(s)$ for the M/M/1 queue with random arrivals at rate α and exponential service with p.d.f. $\beta e^{-\beta t}$; $t \geqslant 0$. Use the results from our earlier analysis (2.12) plus the note at the end of this chapter to verify that this is correct.

4.6   Solutions to Examples 10

1
$$b(t) = \delta\!\left(t - \frac{1}{\beta}\right) \qquad \therefore \int_0^\infty \delta\!\left(t - \frac{1}{\beta}\right) e^{-st}\, dt = e^{-(s/\beta)}$$

The integrand is zero except for the 'massive spike' at $1/\beta$ which has area 1.
$$\therefore \ B^*(s) = e^{-(s/\beta)}$$

$$\therefore \ V^*(s) = \frac{(1-\rho)s}{s - \alpha + \alpha B^*(s)} = \frac{(\beta - \alpha)s}{\beta(s - \alpha + \alpha e^{-(s/\beta)})}$$

$$= \frac{(\beta - \alpha)s}{\beta\!\left[s - \alpha + \alpha\!\left(1 - \dfrac{s}{\beta} + \dfrac{s^2}{2\beta^2} - \dfrac{s^3}{6\beta^3} + \cdots\right)\right]} = \frac{(\beta - \alpha)s}{\beta s - \alpha s + \dfrac{\alpha s^2}{2\beta} - \dfrac{\alpha s^3}{6\beta^2} + \cdots}$$

$$= \frac{1}{1 + \dfrac{\alpha s}{2\beta(\beta - \alpha)} - \dfrac{\alpha s^2}{6\beta^2(\beta - \alpha)} + \cdots}$$

$$= 1 - \frac{\alpha s}{2\beta(\beta - \alpha)} + \frac{\alpha s^2}{6\beta^2(\beta - \alpha)} + \frac{\alpha^2 s^2}{4\beta^2(\beta - \alpha)^2} + \cdots \text{ up to terms in } s^2$$

$$\therefore \ E[X] = \text{coeff.}(-s) = \frac{\alpha}{2\beta(\beta - \alpha)} = W_q \qquad \text{(Examples 9, question 2)}$$

$$E[X^2] = \text{coeff.} \ \frac{s^2}{2} = \frac{\alpha}{3\beta^2(\beta - \alpha)} + \frac{\alpha^2}{2\beta^2(\beta - \alpha)^2}$$

$$\therefore \ \text{Var}[X] = \frac{\alpha}{3\beta^2(\beta - \alpha)} + \frac{\alpha^2}{2\beta^2(\beta - \alpha)^2} - \frac{\alpha^2}{4\beta^2(\beta - \alpha)^2}$$

$$= \frac{4\alpha(\beta - \alpha) + 6\alpha^2 - 3\alpha^2}{12\beta^2(\beta - \alpha)^2} = \frac{4\alpha\beta - \alpha^2}{12\beta^2(\beta - \alpha)^2}$$

2 From (4.12)

$$V^*(s) = \frac{(1-\rho)s}{s - \alpha + \alpha B^*(s)} \qquad \left(\rho = \frac{\alpha}{\beta}\right)$$

and

$$B^*(s) = \int_0^\infty \beta e^{-\beta t} e^{-st}\, dt = \frac{\beta}{\beta + s}$$

$$\therefore \; V^*(s) = \frac{(\beta - \alpha)s}{\beta\left[s - \alpha + \dfrac{\alpha\beta}{\beta + s}\right]} = \frac{(\beta - \alpha)s(\beta + s)}{\beta[\alpha\beta + (s - \alpha)(s + \beta)]}$$

$$= \frac{(\beta - \alpha)(\beta + s)s}{\beta[s^2 - (\alpha - \beta)s]} = \frac{(\beta - \alpha)(\beta + s)}{\beta(s - \alpha + \beta)}$$

From the note following (2.12), $\Pr(X = 0) = 1 - \alpha/\beta$,

$$\varphi(x) = \alpha\left(1 - \frac{\alpha}{\beta}\right)e^{-(\beta - \alpha)x} \qquad \text{with Laplace transform}$$

$$\frac{\alpha}{\beta}(\beta - \alpha)\int_0^\infty e^{-(s + \beta - \alpha)x}\, dx = \frac{\alpha}{\beta}\frac{(\beta - \alpha)}{(s + \beta - \alpha)}$$

Therefore, from these results

$$V^*(s) \;\; = \frac{\beta - \alpha}{\beta} + \frac{\alpha(\beta - \alpha)}{\beta(s + \beta - \alpha)} = \frac{(\beta - \alpha)}{\beta(s + \beta - \alpha)}[s + \beta - \alpha + \alpha]$$

$$\therefore \; V^*(s) = \frac{(\beta - \alpha)(\beta + s)}{\beta(s + \beta - \alpha)} \qquad \text{as before}$$

Note we cannot neglect the discrete part of the distribution

$$V^*(s) = \mathrm{E}[e^{-sX}] = \Pr(X = 0)e^{-s0} + \int_0^\infty \varphi(x)e^{-sx}\, dx$$

5
The imbedded Markov process

5.1 Single-server queue, random arrivals at rate a, general service time – M/G/1

In this system the instants of service completion are points of regeneration of the process. The future of the process is wholly determined by the situation at these instants. Whatever happened before has no bearing. We can describe this situation by the number of customers in the system at the instants of service completion. We have already looked at this problem in Chapter 4. Here we take a slightly different view to derive the steady-state situation.

$$\text{Pr(state } k) = \underset{n \to \infty}{\text{Limit}} \ \text{Pr}(q_n = k) = \pi_k \tag{5.1}$$

Now for random arrivals at rate a, the probability that exactly j customers arrive during a service time (which has p.d.f. $b(t)$; $t \geqslant 0$) is

$$\text{Pr}(\zeta_n = j) = r_j = \int_0^\infty \frac{e^{-at}(at)^j}{j!} \ b(t) \ dt$$

[Refer back to (4.9) and its derivation.]

The r_j give the probabilities of the various transitions that can occur. We relate (in a probability sense) the situation when customer $(n + 1)$ leaves to that which obtained when customer n left ($n \to \infty$). Thus

$$
\begin{aligned}
\pi_0 &= \pi_0 r_0 + \pi_1 r_0 \\
\pi_1 &= \pi_0 r_1 + \pi_1 r_1 + \pi_2 r_0 \\
\pi_2 &= \pi_0 r_2 + \pi_1 r_2 + \pi_2 r_1 + \pi_3 r_0 \\
\pi_3 &= \pi_0 r_3 + \pi_1 r_3 + \pi_2 r_2 + \pi_3 r_1 + \pi_4 r_0
\end{aligned}
\tag{5.2}
$$

$$\text{etc.}$$

The first column on the right in these equations refers to the case where the system is left empty with probability π_0. The next customer to arrive finds the system empty and goes straight into service. That customer will leave behind the number to arrive during their service time.

For the other columns, the number left behind by a customer is the number left behind by the previous customer plus the number to arrive during the next service time minus one (the customer in question). [Refer back to (4.1) and its derivation.] It is these mutually exclusive alternatives which give rise to the sums of (products of) probabilities in the rows of (5.2).

We have to solve these equations for π_n. We can of course calculate r_n. We define the probability generating functions

$$Q(z) = \sum_{k=0}^{\infty} \pi_k z^k$$

and

$$\Phi(z) = \sum_{k=0}^{\infty} r_k z^k = \sum_{k=0}^{\infty} z^k \int_0^{\infty} \frac{e^{-at}(at)^k}{k!} b(t) \, dt = B^*(a(1-z)) \qquad \text{(from (4.9))}$$

Thus we can calculate $\Phi(z)$ if we know $b(t)$. Then multiplication of the successive equations of (5.2) by z^0, z^1, z^2, etc. followed by addition gives (the terms arise column by column)

$$Q(z) = \pi_0 \Phi(z) + \pi_1 \Phi(z) + \pi_2 z \Phi(z) + \pi_3 z^2 \Phi(z) + \cdots$$

$$= (\pi_0 + \pi_1 + \pi_2 z + \pi_3 z^2 + \cdots) \Phi(z)$$

$$= \left(\pi_0 + \frac{Q(z) - \pi_0}{z}\right) \Phi(z)$$

Thus

$$Q(z) = \frac{\pi_0(1-z)\Phi(z)}{\Phi(z) - z}$$

We can deduce the value of π_0 since $Q(1) = \pi_0 + \pi_1 + \pi_2 + \cdots = 1$. Of course we get an indeterminate form $(0/0)$ on the right when $z = 1$. Thus, on using l'Hopital's rule

$$1 = \frac{-\pi_0}{\Phi'(1) - 1}$$

Thus

$$\pi_0 = 1 - \Phi'(1)$$

$$= 1 - \int_0^{\infty} e^{-at} b(t) \left(\sum_{k=1}^{\infty} \frac{(at)^k}{(k-1)!}\right) dt$$

$$= 1 - \int_0^{\infty} at b(t) \, dt = 1 - aE[T]$$

$$\pi_0 = 1 - \rho \qquad \qquad \text{[where } \rho = a \times \text{mean service time]}$$

Thus

$$Q(z) = \frac{(1-\rho)(1-z)\Phi(z)}{\Phi(z) - z} = \frac{(1-\rho)(1-z)B^*[a(1-z)]}{B^*[a(1-z)] - z} \qquad (5.3)$$

a result which we obtained earlier, i.e. (4.8), by a different method.

Example 5.1

Suppose we have random arrivals at rate 1 every 12 minutes. Suppose service time is made up of two independent stages each of which has a negative exponential distribution with mean 1 minute. Find the probability that there are n customers in the system and find the mean number of customers in the system.

Here $\alpha = 1/12$ arrivals/minute. The mean service time $= 1 + 1 = 2$ minutes. Thus $\rho = 2/12 = 1/6$. The service time $T = T_1 + T_2$ where T_i has a negative exponential distribution with p.d.f. $1.e^{-t}$; $t \geq 0$ and Laplace transform

$$\int_0^\infty e^{-t} e^{-st} \, dt = \frac{1}{1+s}$$

Thus T has Laplace transform (from (1.7) since T_1 and T_2 are independent)

$$\left(\frac{1}{1+s}\right)^2 = B^*(s)$$

$\therefore$ $\Phi(z)$ (in the previous notation) $= \left[\dfrac{1}{1 + \dfrac{1}{12}(1-z)}\right]^2 = \dfrac{144}{(13-z)^2}$

Thus

$$Q(z) = \frac{(1-\rho)(1-z)\dfrac{144}{(13-z)^2}}{\dfrac{144}{(13-z)^2} - z}$$

$$= \frac{120(1-z)}{144 - z(13-z)^2} = \frac{120(1-z)}{144 - 169z + 26z^2 - z^3}$$

$$= \frac{120(1-z)}{(1-z)(144 - 25z + z^2)} = \frac{120}{144 - 25z + z^2}$$

[Note $Q(1) = 1$; just a check.]

$$\therefore Q(z) = \sum_{k=0}^{\infty} \pi_k z^k = \frac{120}{(z-9)(z-16)}$$

$$\pi_0 = Q(0) = \frac{120}{144} = \frac{5}{6} \quad (= 1 - \rho)$$

$$Q(z) = \frac{120}{7}\left[\frac{1}{9-z} - \frac{1}{16-z}\right]$$

$$= \frac{120}{7}\left\{\frac{1}{9}\left(1 - \frac{z}{9}\right)^{-1} - \frac{1}{16}\left(1 - \frac{z}{16}\right)^{-1}\right\}$$

$$= \frac{120}{7}\left\{\frac{1}{9}\left(1 + \frac{z}{9} + \frac{z^2}{9^2} + \cdots\right) - \frac{1}{16}\left(1 + \frac{z}{16} + \frac{z^2}{16^2} + \cdots\right)\right\}$$

$$\therefore \ \pi_n = \text{coeff. of } z^n$$

$$\therefore \ \pi_n = \frac{120}{7}\left[\left(\frac{1}{9}\right)^{n+1} - \left(\frac{1}{16}\right)^{n+1}\right]$$

The mean number of customers in the system is given by

$$Q'(1) = \sum_{k=0}^{\infty} k\pi_k$$

$$Q'(z) = \frac{120}{7}\left[\frac{1}{(9-z)^2} - \frac{1}{(16-z)^2}\right]$$

$$\therefore \ Q'(1) = \frac{120}{7}\left[\frac{1}{64} - \frac{1}{225}\right] = \frac{120 \times 161}{7 \times 64 \times 225}$$

$$= \frac{120 \times 23}{64 \times 225} = \frac{23}{8 \times 15} = \frac{23}{120}$$

5.2 Examples 11

1 Casualties arrive at the accident reception centre of a hospital at a rate of 1 every 5 hours. Treatment consists of two independent phases: (i) X-ray and examination which takes 15 minutes, and (ii) time in the operating theatre which has mean 1 hour and standard deviation $1/2$ hour. A criterion to be satisfied is that the average waiting time from admission to the start of treatment should not exceed 15 minutes. Find whether this criterion will be satisfied.

2 For a single-server queue customers arrive at random at rate α and the length of service has an exponential distribution with parameter β. Show that the probability that j customers arrive during a service time is

$$r_j = \left(\frac{\alpha}{\alpha+\beta}\right)^j \cdot \frac{\beta}{\alpha+\beta}; \qquad j = 0,1,2,\ldots$$

Deduce that

$$\Phi(z) = \sum_{j=0}^{\infty} r_j z^j = \frac{\beta}{\beta+\alpha(1-z)}$$

Let the probability that there are k in the system when a service finishes be π_k and let

$$Q(z) = \sum_{k=0}^{\infty} \pi_k z^k$$

Use the result $Q(z) = (\pi_0(1-z)\Phi(z))/(\Phi(z)-z)$ and deduce that

$$\pi_n = \left(1 - \frac{\alpha}{\beta}\right)\left(\frac{\alpha}{\beta}\right)^n$$

3 A quality control inspector inspects incoming batches of raw materials. These arrive at random at rate 5/hour. The inspector carries out two independent tests on each batch. The time to carry out each test has a negative exponential distribution with mean 2 minutes.

Show that π_n, the probability that there are n batches either being inspected or waiting to be inspected, is given by

$$\pi_n = \frac{24}{5}\left\{\left(\frac{1}{4}\right)^{n+1} - \left(\frac{1}{9}\right)^{n+1}\right\}$$

Find the mean number of batches either waiting or being inspected.

5.3   Solutions to examples 11

1 $\alpha = 1/5$, $\mathrm{E}[T] = 1/4 + 1 = 5/4$, $\mathrm{Var}[T] = 1/4$, $\rho = \alpha\mathrm{E}[T] = 1/4$. Note that we are assuming random arrivals.

$$\therefore \ \alpha W_q = L_q = \frac{\rho^2 + \alpha^2 \, \mathrm{Var}[T]}{2(1 - \rho)}$$

$$\therefore \ \frac{W_q}{5} = \frac{\dfrac{1}{16} + \dfrac{1}{25}\dfrac{1}{4}}{2\left(1 - \dfrac{1}{4}\right)}$$

$$\therefore \ W_q = \frac{\dfrac{5}{16} + \dfrac{1}{20}}{\dfrac{3}{2}} = \frac{2}{3}\left(\frac{25 + 4}{80}\right) = \frac{29}{120} \text{ hours} \simeq \frac{1}{4}$$

The criterion is just satisfied.

2
$$r_j = \int_0^\infty \frac{(\alpha t)^j e^{-\alpha t} \beta e^{-\beta t}}{j!} \, dt = \int_0^\infty \frac{\beta(\alpha t)^j e^{-(\alpha + \beta)t}}{j!} \, dt$$

Put $v = (\alpha + \beta)t$

$$r_j = \int_0^\infty \frac{\beta}{\alpha + \beta} \frac{\alpha^j}{(\alpha + \beta)^j} \frac{v^j e^{-v}}{j!} \, dv = \left(\frac{\alpha}{\alpha + \beta}\right)^j \frac{\beta}{\alpha + \beta}$$

$$\Phi(z) = \sum_{j=0}^\infty r_j z^j = \frac{\beta}{\alpha + \beta} \sum_{j=0}^\infty \left(\frac{z\alpha}{\alpha + \beta}\right)^j = \frac{\beta}{\alpha + \beta} \frac{1}{1 - \dfrac{z\alpha}{\alpha + \beta}} = \frac{\beta}{\beta + \alpha(1 - z)}$$

$$Q(z) = \frac{\pi_0(1-z)\dfrac{\beta}{\beta+\alpha(1-z)}}{\dfrac{\beta}{\beta+\alpha(1-z)} - z} = \frac{\pi_0\beta(1-z)}{\beta - (\beta+\alpha)z + \alpha z^2} = \frac{\pi_0\beta(1-z)}{(1-z)(\beta - \alpha z)}$$

$$\therefore\ Q(z) = \frac{\pi_0\beta}{(\beta - \alpha z)} = \frac{\pi_0}{\left(1 - \dfrac{\alpha}{\beta}z\right)}$$

Since $Q(1) = 1$, $\pi_0 = 1 - \alpha/\beta$.

$$\therefore\ Q(z) = \left(1 - \frac{\alpha}{\beta}\right)\left(1 + \frac{\alpha}{\beta}z + \left(\frac{\alpha}{\beta}\right)^2 z^2 + \left(\frac{\alpha}{\beta}\right)^3 z^3 + \cdots\right)$$

$$\pi_k = \left(1 - \frac{\alpha}{\beta}\right)\left(\frac{\alpha}{\beta}\right)^k$$

3 With minutes as the time units $\alpha = 1/12$. Mean service time $= E[T] = 2 + 2 = 4$, $\rho = 4/12 = 1/3$. The service time $T = T_1 + T_2$ where each T_i has p.d.f. $1/2e^{-t/2}$; $t \geqslant 0$ and Laplace transform

$$\int_0^\infty \frac{1}{2}e^{-t/2}e^{-st}\,dt = \frac{1}{2}\bigg/\left(\frac{1}{2} + s\right) = \frac{1}{1+2s}$$

Thus T has Laplace transform $(1/(1+2s))^2 = B^*(s)$.

$$\therefore\ \Phi(z) = \left[\frac{1}{1 + \dfrac{1}{6}(1-z)}\right]^2 = \frac{36}{(7-z)^2}$$

$$\therefore\ Q(z) = \frac{\dfrac{2}{3}(1-z)\dfrac{36}{(7-z)^2}}{\dfrac{36}{(7-z)^2} - z} = \frac{24(1-z)}{36 - 49z + 14z^2 - z^3}$$

$$\therefore\ Q(z) = \frac{24(1-z)}{(1-z)(36 - 13z + z^2)} = \frac{24}{(9-z)(4-z)}$$

$$= \frac{24}{5}\left[\frac{1}{4-z} - \frac{1}{9-z}\right]$$

$$= \frac{24}{5}\left\{\frac{1}{4}\left(1 + \frac{z}{4} + \frac{z^2}{4^2} + \cdots\right) - \frac{1}{9}\left(1 + \frac{z}{9} + \frac{z^2}{9^2} + \frac{z^3}{9^3} + \cdots\right)\right\}$$

$$\therefore \; \pi_n = \frac{24}{5} \left\{ \left(\frac{1}{4}\right)^{n+1} - \left(\frac{1}{9}\right)^{n+1} \right\}$$

$$E[N] = Q'(1) = \frac{24}{5} \left[\frac{1}{(4-z)^2} - \frac{1}{(9-z)^2} \right]\Bigg|_{z=1} = \frac{24}{5} \left[\frac{1}{9} - \frac{1}{64} \right]$$

$$E[N] = \frac{24}{5} \times \frac{55}{9 \times 64} = \frac{11}{24}$$

5.4   Single-server queue, general inter-arrival time, exponential service – G/M/1

Inter-arrival times are independent, identically distributed with p.d.f. $a(t)$; $t \geqslant 0$. Service times have a negative exponential distribution with p.d.f. $\beta e^{-\beta t}$. We consider those instants at which a new customer arrives. Let customer n find R_n customers in the system on arrival. As usual we consider the steady-state situation which is assumed to exist; for example

$$\underset{n \to \infty}{\text{Limit}} \; \Pr(R_n = k) = u_k \tag{5.4}$$

Let v_j denote the probability that exactly j customers are served in an inter-arrival time during which service is ongoing;

$$v_j = \int_0^\infty \frac{e^{-\beta t}(\beta t)^j}{j!} \, a(t) \, \mathrm{d}t$$

$t_j = 1 - \sum_{i=0}^j v_i$ is used to denote the probability that more than j customers can be served in an inter-arrival time. Then

$$u_0 = u_0 t_0 + u_1 t_1 + u_2 t_2 + \cdots$$
$$u_1 = u_0 v_0 + u_1 v_1 + u_2 v_2 + \cdots$$
$$u_2 = u_1 v_0 + u_2 v_1 + u_3 v_2 + \cdots \tag{5.5}$$
$$u_3 = u_2 v_0 + u_3 v_1 + u_4 v_2 + \cdots$$
$$\text{etc.}$$

i.e.

$$u_k = \sum_{i=0}^\infty u_{i+k-1} v_i, \qquad k = 1, 2, 3, \ldots$$

In fact it can be shown that the first equation really says nothing new that is not capable of deduction from the others (see question 3 of Examples 12). For the other equations, the L.H.S. gives the probability that customer $(n+1)$ finds k in the system. For this to happen customer n must have found $i+k-1$ in the system (so that with customer n there were $k+i$ in the system immediately after the arrival) and there were i service completions during the inter-arrival time. The R.H.S. consists of the sum of the probabilities corresponding to these mutually exclusive possibilities.

A solution to these equations exists which is of the form

$$u_k = A + B\eta^k$$

where A, B and η are constants to be found. For the equation with $k = 1$, direct substitution gives

$$A + B\eta = \sum_{i=0}^{\infty} Av_i + B \sum_{i=0}^{\infty} v_i \eta^i$$
$$= A + BP(\eta)$$

where $P(z) = \sum_{i=0}^{\infty} v_i z^i$ is the p.g.f. for the v_i. For general k, substitution gives

$$A + B\eta^k = A \sum_{i=0}^{\infty} v_i + B\eta^{k-1} \sum_{i=0}^{\infty} v_i \eta^i$$

Thus $A + B\eta^k$ is a solution provided $\eta = P(\eta)$.

Since $\sum_{k=0}^{\infty} u_k = 1$ we must have $A = 0$ and $\eta < 1$ so that the series is convergent. Thus $u_0 = B$, $u_1 = B\eta$, $u_2 = B\eta^2, \dots$ and so $B = (1 - \eta)$. Thus

$$u_k = (1 - \eta)\eta^k \qquad (5.6)$$

where $\eta < 1$ is the smallest root of the equation $z = P(z)$. For such a root to exist $P'(1) > 1$. Now $P'(1) = \sum_{i=0}^{\infty} iv_i$ is the mean number of services completed in an inter-arrival time. This must exceed 1 for a steady state to exist. This is clear from Fig. 5.1 and is as expected intuitively. The service rate exceeds the arrival rate. Then

$$u_k = \Pr(R_n = k) = (1 - \eta)\eta^k \qquad \text{where } \eta = P(\eta) \text{ and } \eta < 1$$

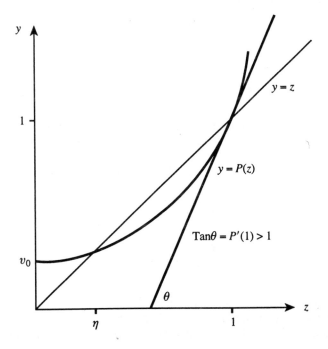

Fig. 5.1

For negative exponential service time the number of customers found in the system by an arrival has a geometric distribution. $u_0 = (1 - \eta)$ is the probability that the system is found (or left) empty. Of course

$$P(z) = \sum_{j=0}^{\infty} z^j v_j = \sum_{j=0}^{\infty} z^j \int_0^{\infty} \frac{(\beta t)^j e^{-\beta t}}{j!} \, a(t) \, \mathrm{d}t$$

$$= \int_0^{\infty} a(t) e^{-\beta(1-z)t} \, \mathrm{d}t = A^*(\beta(1-z)) \tag{5.7}$$

where $A^*(s) = \int_0^{\infty} e^{-st} a(t) \, \mathrm{d}t$ is the Laplace transform of $a(t)$.

If we consider the time X spent in the queue by an arrival in the steady-state situation

$$\Pr(X = 0) = u_0 = 1 - \eta \tag{5.8}$$

When X is non-zero it has a continuous distribution with p.d.f. $\varphi(x)$ given by (by the same argument as led to (2.12)),

$$\varphi(x) = \sum_{j=1}^{\infty} u_j \frac{(\beta x)^{j-1} e^{-\beta x}}{(j-1)!} \beta$$

$$= \sum_{j=1}^{\infty} \frac{(1-\eta)\eta^j (\beta x)^{j-1} e^{-\beta x}}{(j-1)!} \beta$$

$$= \beta \eta (1 - \eta) e^{-\beta x(1-\eta)}; \qquad x > 0 \tag{5.9}$$

The time spent in the system is $Y = X + S$ where S is the time in service. Y is a continuous random variable with p.d.f.

$$\psi(y) = \beta(1 - \eta) e^{-\beta(1-\eta)y}; \qquad y \geq 0$$

For

$$\psi(y) = \Pr(X = 0)\beta e^{-\beta y} + \int_0^y \varphi(x)\beta e^{-\beta(y-x)} \, \mathrm{d}x$$

| zero queueing service time y | queueing time x | service time $y-x$ |

$$\therefore \psi(y) = (1 - \eta)\beta e^{-\beta y} + \int_0^y \beta\eta(1 - \eta) e^{-\beta x(1-\eta)} \beta e^{-\beta(y-x)} \, \mathrm{d}x$$

$$= (1 - \eta)\beta e^{-\beta y} + \beta^2 \eta(1 - \eta) e^{-\beta y} \int_0^y e^{\beta \eta x} \, \mathrm{d}x$$

$$= (1 - \eta)\beta e^{-\beta y} + \beta(1 - \eta) e^{-\beta y}(e^{\beta \eta y} - 1)$$

$$= \beta(1 - \eta) e^{-\beta(1-\eta)y}; \qquad y \geq 0 \tag{5.10}$$

Thus if service has a negative exponential distribution with parameter β, time in the system has a negative exponential distribution with parameter $\beta(1 - \eta)$. The negative exponential nature of this latter distribution is so whatever the inter-arrival time distribution.

5.5 Examples 12

1 For the system with one server, independent, identically distributed inter-arrival times and negative exponential service at rate β, show that

(a) the mean number of customers found in the system by an arrival is
$L = \eta/(1 - \eta)$
(b) the mean time spent in the queue is $\eta/\beta(1 - \eta)$
(c) the mean time spent in the system is $1/(\beta(1 - \eta))$

where η (<1) is the smallest root of $z = A^*\{\beta(1 - z)\}$ and $A^*(s)$ is the Laplace transform of the inter-arrival time p.d.f.

In the case of random arrivals at rate α show that $\eta = \alpha/\beta$ so that results (a), (b), (c) above are in accord with those for the M/M/1 system.

2 With one server and random arrivals at rate α and negative exponential service at rate β we have seen ((2.10) and above) that the mean number in the system is $L = \rho/(1 - \rho)$ where $\rho = \alpha/\beta$ (<1). Suppose the inter-arrival time is constant at $1/\alpha$ (regular arrivals at the same average rate as before).
Show that η (above) is the smallest root of the equation

$$z = e^{-\beta/\alpha(1 - z)} = e^{-(1 - z)/\rho}$$

Show that if $\rho < 1$ the smallest root of the equation $z = \exp\{-(1 - z)/\rho\}$ is less than ρ.
Use (a) of question 1 to deduce that, in terms of the average number in the system, regular arrivals are better than random arrivals at the same average rate. [This is why your doctor, etc. should run an appointments system.]

3 Consider the first of the equations in (5.5). Write $t_0 = 1 - v_0$, $t_1 = 1 - v_0 - v_1$, etc. and hence show that the R.H.S. is u_0. Thus this equation can be deduced from the others.

5.6 Solutions to examples 12

1 (a) $L = \Sigma k u_k = \displaystyle\sum_{k=0}^{\infty} (1 - \eta)k\eta^k = \dfrac{\eta}{1 - \eta}$ $\left[\text{N.B. } \Sigma nx^n = \dfrac{x}{(1 - x)^2} \text{ for } |x| < 1\right]$

(b) $W_q = 0\Pr(X = 0) + \displaystyle\int_0^{\infty} \beta\eta(1 - \eta)xe^{-\beta x(1 - \eta)} \, dx$

$= \dfrac{\beta\eta(1 - \eta)}{[\beta(1 - \eta)]^2} = \dfrac{\eta}{\beta(1 - \eta)}$

(c) $W = W_q + \dfrac{1}{\beta} = \dfrac{\eta}{\beta(1 - \eta)} + \dfrac{1}{\beta} = \dfrac{1}{\beta(1 - \eta)}$

or $W = \displaystyle\int_0^{\infty} \beta(1 - \eta)ye^{-\beta(1 - \eta)y} \, dy = \dfrac{1}{\beta(1 - \eta)}$

If $a(t) = \alpha e^{-\alpha t}$

$$A^*(s) = \int_0^\infty \alpha e^{-(\alpha + s)t} \, dt = \frac{\alpha}{\alpha + s}$$

$$\therefore \ z = A^*[\beta(1 - z)] \text{ is } z = \frac{\alpha}{\alpha + \beta(1 - z)}$$

$$\therefore \ \beta z^2 - (\alpha + \beta)z + \alpha = 0 \qquad \therefore \ (z - 1)(\beta z - \alpha) = 0$$

Therefore, $z = 1$ or $z = \alpha/\beta$ and so $\eta = \alpha/\beta$ (the smallest root <1).

2 If the inter-arrival time is constant at $1/\alpha$, $a(t) = \delta(t - 1/\alpha)$ and $A^*(s) = e^{-s/\alpha}$ [see Examples 10, question 1]. Thus

$$A^*(\beta(1 - z)) = e^{-\beta(1 - z)\alpha} = e^{-(1 - z)/\rho}$$

where $\rho = \alpha/\beta$. The equation to be solved is $z = \exp\{-(1 - z)/\rho\}$ where $\rho < 1$ (Fig. 5.2). We must have

$$1 > \tan \theta = P'(\eta) = \frac{1}{\rho} e^{-1/\rho(1 - \eta)} = \frac{\eta}{\rho}$$

since $\eta = \exp\{-1/\rho(1 - \eta)\}$

$$\therefore \ \frac{\eta}{\rho} < 1$$

$$\therefore \ \eta < \rho < 1$$

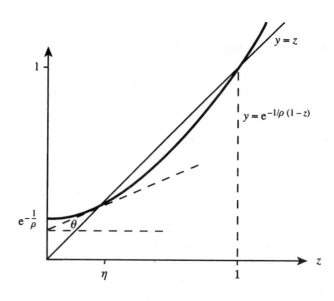

Fig. 5.2

Now if $\eta < \rho < 1$

$$\therefore \quad \frac{1}{1-\eta} < \frac{1}{1-\rho}$$

$$\therefore \quad \frac{\eta}{1-\eta} < \frac{\rho}{1-\rho} \qquad \text{as required}$$

Thus a regular arrivals system D/M/1 [D for deterministic] reduces the average number in the system. It also decreases the waiting time in the queue and in the system.

3 Since $t_j = 1 - v_0 - v_1 - v_2 - \cdots - v_j$ the R.H.S. of the equation which is

$$u_0 t_0 + u_1 t_1 + u_2 t_2 + u_3 t_3 + \cdots$$
$$= u_0(1 - v_0)$$
$$\quad + u_1(1 - v_0 - v_1)$$
$$\quad + u_2(1 - v_0 - v_1 - v_2)$$
$$\quad + u_3(1 - v_0 - v_1 - v_2 - v_3)$$
$$\text{etc.}$$
$$= u_0 + u_1 + u_2 + u_3 + \cdots$$
$$\quad - (u_0 v_0 + u_1 v_1 + u_2 v_2 + u_3 v_3 + \cdots)$$
$$\quad - (u_1 v_0 + u_2 v_1 + u_3 v_2 + \cdots)$$
$$\quad - (u_2 v_0 + u_3 v_1 + \cdots)$$
$$\quad - (u_3 v_0 + \cdots)$$
$$\text{etc.}$$

But from the later equations of the set (5.5) the bracketed expressions are $u_1, u_2, u_3, \ldots$, etc. in turn. Thus the R.H.S. of the first equation in the set (5.5)

$$= u_0 + u_1 + u_2 + u_3 + u_4 + \cdots$$
$$\quad - u_1 - u_2 - u_3 - u_4 - \cdots$$
$$= u_0$$

Thus this first equation contains no more information about the system. This explains why we did not need to use it to obtain the solution (5.6).

5.7 The Markov chain formulation

Both of the previous models M/G/1 and G/M/1 can be thought of as Markov chains with states described by the number of customers left in the system by a departing customer or the number of customers found in the system by a new arrival, respectively. Changes in the state at successive departure or arrival times are brought about via the number of arrivals during a service time or the number of departures during an inter-arrival time, and we construct the transition probability matrices appropriate to the two situations.

The M/G/1 system

Consider the steady-state situation. We take a second look at the first system investigated in Chapters 4 and 5 and show explicitly the imbedded Markov chain. Let

$X_r \equiv$ number of customers left behind by customer r at his moment of departure ($\equiv q_r$ of Chapter 4), and $Z_r \equiv$ number of customers who arrive during the service time of customer r ($\equiv \xi_r$ of Chapter 4). Then

$$X_{r+1} = \begin{cases} X_r - 1 + Z_{r+1}, & X_r \geqslant 1 \\ Z_{r+1}, & X_r = 0 \end{cases} \tag{5.11}$$

Of course (5.11) is equivalent to (4.1).

The transition probability $p_{ij} = \Pr(X_{r+1} = j \,|\, X_r = i)$

$$p_{0j} = \int_0^\infty \frac{(at)^j}{j!}\, e^{-at} b(t)\, dt = r_j \text{ say} \qquad [j \text{ arrivals during the service time}]$$

$$p_{ij} = \int_0^\infty \frac{(at)^{j-i+1}}{(j-i+1)!}\, e^{-at} b(t)\, dt = r_{j-i+1} \text{ say} \qquad [j - i + 1 \text{ arrivals}]$$

$$p_{ij} = 0 \qquad \text{for } j < i - 1$$

Thus the one-step transition matrix takes the form

$$\mathbf{P} = \begin{array}{c} \\ 0 \\ 1 \\ 2 \\ 3 \\ 4 \\ \\ \end{array}\!\!\begin{array}{c} \begin{array}{cccccc} 0 & 1 & 2 & 3 & & \end{array} \\ \left(\begin{array}{cccccc} r_0 & r_1 & r_2 & r_3 & \cdots \\ r_0 & r_1 & r_2 & r_3 & \cdots \\ 0 & r_0 & r_1 & r_2 & \cdots \\ 0 & 0 & r_0 & r_1 & \cdots \\ 0 & 0 & 0 & r_0 & \cdots \\ & \vdots & & & \end{array}\right) \end{array}$$

This is the transition matrix for an irreducible Markov chain since all states communicate. All states are aperiodic. Thus the chain is ergodic and there is an equilibrium (steady-state) distribution π which is the unique solution of

$$\pi = \pi \mathbf{P} \qquad \text{where } \pi = (\pi_0, \pi_1, \pi_2, \ldots).$$

Thus the equations for π are

$$\pi_0 = \pi_0 r_0 + \pi_1 r_0$$
$$\pi_1 = \pi_0 r_1 + \pi_1 r_1 + \pi_2 r_0$$
$$\pi_2 = \pi_0 r_2 + \pi_1 r_2 + \pi_2 r_1 + \pi_3 r_0$$
$$\text{etc.}$$

These are the equations (5.2) and can be solved in the manner shown there using generating functions.

The G/M/1 system

Consider the steady-state situation. We take a second look at the other system investigated in 5.4 and show explicitly the imbedded Markov chain. Let $X_r \equiv$ the number of customers ahead of him found by customer r on arrival, and $Y_r \equiv$ the number of departures between the arrival of the rth and $(r+1)$th customers.

The distribution of Y_r depends on the service rate β, the inter-arrival time distribution $a(t)$, and on X_r, but not on earlier values of X. Thus we have a Markov chain for the number in the system at the instants of arrival with transition probability

$$p_{ij} = \Pr(X_{r+1} = j \mid X_r = i) = \Pr(Y_r = i + 1 - j \mid X_r = i)$$

$$\therefore p_{ij} = \int_0^\infty \frac{(\beta t)^{i+1-j}}{(i+1-j)!} e^{-\beta t} a(t) \, dt = v_{i+1-j} \qquad j \leqslant i+1$$

$$= 0 \qquad \text{for } j > i + 1$$

When $j \neq 0$ the server remains busy throughout; hence the above. For $j = 0$ we want the probability that the server clears the queue, so that he would serve at least $(i + 1)$ customers if he could go on serving.

$$p_{i0} = \int_0^\infty \sum_{k=i+1}^\infty \frac{(\beta t)^k e^{-\beta t}}{k!} a(t) \, dt = \sum_{k=i+1}^\infty v_k = 1 - \sum_{k=0}^i v_k = t_i$$

Thus the one-step transition matrix is

$$\mathbf{P} = \begin{array}{c} \\ 0 \\ 1 \\ 2 \\ 3 \end{array} \begin{array}{ccccc} 0 & 1 & 2 & 3 & 4 \\ \left(\begin{array}{ccccc} t_0 & v_0 & 0 & 0 & 0 & \cdots \\ t_1 & v_1 & v_0 & 0 & 0 & \cdots \\ t_2 & v_2 & v_1 & v_0 & 0 & \cdots \\ t_3 & v_3 & v_2 & v_1 & v_0 & \cdots \end{array}\right) \end{array}$$

From the form of $\mathbf{P}$ (all $v_k > 0$) the Markov chain is irreducible and aperiodic and hence ergodic. For the stationary distribution u,

$$u = u\mathbf{P}$$

These are equations (5.5). The first equation is

$$\therefore u_0 = \sum_{i=0}^\infty u_i t_i = \sum_{i=0}^\infty u_i \sum_{k=i+1}^\infty v_k = \sum_{k=1}^\infty v_k \sum_{i=0}^{k-1} u_i$$

on changing the order of summation. The later equations are of the form

$$u_k = \sum_{i=k-1}^\infty u_i v_{i+1-k} = \sum_{i=0}^\infty u_{i+k-1} v_i \qquad \text{for } k \geqslant 1$$

If we try the solution $u_k = (1 - \eta)\eta^k$, $k = 0, 1, 2, \ldots$, on the R.H.S. we have for the first equation

$$(1 - \eta) \sum_{k=1}^\infty v_k \sum_{i=0}^{k-1} \eta^i$$

$$= \sum_{k=1}^\infty v_k(1 - \eta^k) = \sum_{k=0}^\infty v_k(1 - \eta^k)$$

$$= \left[\sum_{k=0}^\infty v_k - \sum_{k=0}^\infty v_k \eta^k \right]$$

$$= 1 - P(\eta) \qquad \text{where } P(z) = \sum_{k=0}^\infty v_k z^k$$

and for the kth equation

$$(1 - \eta) \sum_{i=0}^{\infty} v_i \eta^{i+k-1} = (1 - \eta)\eta^{k-1}P(\eta)$$

and on the L.H.S., $1 - \eta$ and $(1 - \eta)\eta^k$, respectively. Thus with η a root of $z = P(z)$ the L.H.S. and R.H.S. are equal.

The first equation says nothing other than

$$1 - \eta = 1 - P(\eta)$$

which is no more than that $\eta = P(\eta)$. Thus the first equation does not help us in establishing the solution $u_k = (1 - \eta)\eta^k$ [see question 3 of Examples 12]. This solution $u_k = (1 - \eta)\eta^k$, $k = 0, 1, 2, ...$, will only give a stationary solution if $\eta < 1$. But since $P(z)$ is a p.g.f. with $v_0 > 0$, $P(1) = 1$ and $z = P(z)$ has a root $\eta < 1$ if $P'(1) > 1$. This point has been discussed earlier following (5.6).

Example 5.2

Customers to a single-server queue arrive at random at an average rate of 4/hour. Service time has a negative exponential distribution with mean 12 minutes. Show that $\rho = 0.8$ and that the average number of customers in the system is 4 and the mean queueing time for a customer is 48 minutes. Suppose an appointments system is organised so that customers arrive at 15-minute intervals. Show that, in the notation of the notes, $\eta \approx 0.62863$ ($< \rho$ of course).

Deduce that the average number of customers in the system when a customer arrives is 1.693 and the mean waiting time in the queue is 20.31 minutes. What is the probability that a customer has to wait in the case of (i) random arrivals, (ii) regular arrivals?

With our time unit as 1 minute the arrival rate $\alpha = 1/15$ and the service rate $\beta = 1/12$. Thus in the notation for an M/M/1 queue $\rho = \alpha/\beta = 12/15 = 0.8$. The average number in the system, from (2.10), is

$$L = \frac{\rho}{1 - \rho} = \frac{0.8}{0.2} = 4$$

The mean queueing time for a customer, from (2.13), is

$$W_q = \frac{\alpha}{\beta(\beta - \alpha)} = \frac{0.8}{\dfrac{1}{12} - \dfrac{1}{15}} = \frac{0.8 \times 120}{2}$$

$$= 48 \text{ minutes}$$

If the inter-arrival time is constant, 15 minutes, then the probability density function $a(t) = \delta(t - 15)$ [see Fig. 5.3].

$$\therefore \ A^*(s) = \int_0^{\infty} e^{-st} \delta(t - 15) \, dt = e^{-15s}$$

$$\therefore \ P(z) = A^*[\beta(1 - z)] = e^{-5(1-z)/4}$$

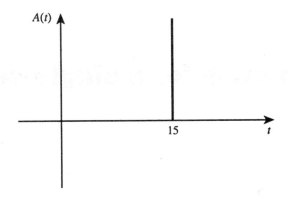

Fig. 5.3

Thus we find η from $z - \exp\{-5(1-z)/4\} = 0$ and using an appropriate numerical method, $\eta = = 0.62863$. Then

$$L' = \frac{\eta}{1-\eta} = 1.693 \quad \text{and} \quad W'_q = \frac{\eta}{\beta(1-\eta)} = 20.31 \text{ minutes}$$

The probability that a customer has to wait is:

(i) for random arrivals, $\rho = (1 - p_0) = 0.8$
(ii) for scheduled arrivals, $\eta = 1 - p'_0 = 0.62863$.

The advantages of scheduling arrivals when this is possible are obvious.

6
Other models for a single-server queue

6.1 Bulk arrivals

Suppose we have a single-server queue with negative exponential service at rate β. Suppose arrivals occur at random at average rate α but each arrival corresponds to a group of k customers. If $p_n(t) \equiv$ probability that there are n customers in the system at time t, then to first order in δt:

$$p_0(t + \delta t) = p_0(t)(1 - \alpha\delta t) + p_1(t)(1 - \alpha\delta t)\beta\delta t$$
$$p_1(t + \delta t) = p_1(t)(1 - \alpha\delta t)(1 - \beta\delta t) + p_2(t)(1 - \alpha\delta t)\beta\delta t$$

$$p_{k-1}(t + \delta t) = p_{k-1}(t)(1 - \alpha\delta t)(1 - \beta\delta t) + p_k(t)(1 - \alpha\delta t)\beta\delta t$$
$$p_k(t + \delta t) = p_0(t)\alpha\delta t + p_k(t)(1 - \alpha\delta t)(1 - \beta\delta t) + p_{k+1}(t)\beta\delta t(1 - \alpha\delta t)$$
$$p_{k+1}(t + \delta t) = p_1(t)\alpha\delta t(1 - \beta\delta t) + p_{k+1}(t)(1 - \alpha\delta t)(1 - \beta\delta t) + p_{k+2}(t)\beta\delta t(1 - \alpha\delta t)$$
$$\text{etc.}$$
$$p_n(t + \delta t) = p_{n-k}(t)\alpha\delta t(1 - \beta\delta t) + p_n(t)(1 - \alpha\delta t)(1 - \beta\delta t) + p_{n+1}\beta\delta t(1 - \alpha\delta t)$$

The arguments involved here are precisely those involved in the discussion leading to (2.4) or (3.1). We consider the interval $(t, t + \delta t)$ and the events which must occur within that interval in order to obtain the event described on the L.H.S.

The first k equations differ from the remaining equations in that an arrival (of k customers) does not occur. Thus if we re-arrange terms and let $\delta t \to 0$ in the usual way

$$\frac{\mathrm{d}p_0(t)}{\mathrm{d}t} = -\alpha p_0(t) + \beta p_1(t)$$

$$\frac{\mathrm{d}p_1(t)}{\mathrm{d}t} = -(\alpha + \beta)p_1(t) + \beta p_2(t)$$

$$\frac{\mathrm{d}p_{k-1}(t)}{\mathrm{d}t} = -(\alpha + \beta)p_{k-1}(t) + \beta p_k(t) \tag{6.1}$$

$$\frac{\mathrm{d}p_k(t)}{\mathrm{d}t} = \alpha p_0(t) - (\alpha + \beta)p_k(t) + \beta p_{k+1}(t)$$

$$\frac{\mathrm{d}p_{k+1}(t)}{\mathrm{d}t} = \alpha p_1(t) - (\alpha + \beta)p_{k+1}(t) + \beta p_{k+2}(t)$$

$$\text{etc.}$$

Of course if $k = 1$ we have the M/M/1 system and (6.1) is identical to (2.4). On the assumption that there is a steady-state solution in which $p_n(t) \to p_n$ and of course $dp_n(t)/dt \to 0$ as $t \to \infty$, then

$$-\alpha p_0 + \beta p_1 = 0$$
$$-(\alpha + \beta)p_1 + \beta p_2 = 0$$

$$-(\alpha + \beta)p_{k-1} + \beta p_k = 0 \qquad (6.2)$$
$$\alpha p_0 - (\alpha + \beta)p_k + \beta p_{k+1} = 0$$
$$\alpha p_1 - (\alpha + \beta)p_{k+1} + \beta p_{k+2} = 0$$
$$\text{etc.}$$

Define the generating function for p_n as

$$P(z) = \sum_{n=0}^{\infty} p_n z^n$$

Then if we multiply equations (6.2) in turn by $z^0, z^1, \ldots, z^{k-1}, z^k, \ldots$ and add we obtain

$$\alpha z^k \sum_{n=0}^{\infty} p_n z^n - \alpha \sum_{n=0}^{\infty} p_n z^n - \beta \sum_{n=1}^{\infty} p_n z^n + \beta \sum_{n=1}^{\infty} p_n z^{n-1} = 0$$

$$\therefore \ \alpha z^k P(z) - \alpha P(z) - \beta(P(z) - p_0) + \frac{\beta(P(z) - p_0)}{z} = 0$$

$$\therefore \ P(z)[\alpha z^{k+1} - (\alpha + \beta)z + \beta] = \beta p_0(1 - z)$$

$$\therefore \ P(z) = \frac{\beta p_0(1 - z)}{\alpha z^{k+1} - (\alpha + \beta)z + \beta} \qquad (6.3)$$

$$P(0) = \frac{\beta p_0}{\beta} = p_0 \qquad \text{as of course it must be}$$

Again we see that if $k = 1$ we can regard (2.7) as a special case of (6.3). Of course we have done a little simplification in obtaining (2.7). We must also have $P(1) = 1$ and this will allow us to evaluate p_0 in terms of α and β.

Simply substituting $z = 1$ into the R.H.S. gives the indeterminate form $0/0$ so we must use l'Hopital's rule;

$$P(1) = \underset{z \to 1}{\text{Limit}} \ \frac{-\beta p_0}{(k+1)\alpha z^k - (\alpha + \beta)} = \frac{-\beta p_0}{k\alpha - \beta} = \frac{\beta p_0}{\beta - k\alpha} = 1$$

Thus

$$p_0 = \frac{\beta - k\alpha}{\beta} = 1 - \frac{k\alpha}{\beta} \qquad (6.4)$$

Of course since p_0 must be positive we see from this result that $k\alpha/\beta < 1$, i.e. $k\rho < 1$ where $\rho = \alpha/\beta$, is a necessary condition for the existence of a steady-state solution. We would expect this. It is equivalent to $\beta > k\alpha$, the service rate of customers is greater than the 'arrival rate of customers'.

$$P(z) = \frac{(1 - k\rho)(1 - z)}{1 - (1 + \rho)z + \rho z^{k+1}} = \frac{(1 - k\rho)(1 - z)}{1 - z - \rho z(1 - z^k)}$$

$$= \frac{(1 - k\rho)}{1 - \rho z \left(\dfrac{1 - z^k}{1 - z} \right)}$$

$$= \frac{(1 - k\rho)}{1 - \rho z(1 + z + z^2 + \cdots + z^{k-1})} = \frac{1 - k\rho}{1 - \rho z - \rho z^2 - \cdots - \rho z^k} \tag{6.5}$$

Certainly we can see that (2.7) for the M/M/1 system is the special case of (6.5) when $k = 1$ and arrivals occur singly.

$$P'(z) = \frac{\mathrm{d}P}{\mathrm{d}z} = \sum_{n=0}^{\infty} n p_n z^{n-1}$$

so that

$$P'(z) \big|_{z=1} = \sum_{n=0}^{\infty} n p_n = \sum_{n=1}^{\infty} n p_n$$

$$P'(z) = \frac{(1 - k\rho)\rho[1 + 2z + 3z^2 + \cdots + kz^{k-1}]}{[1 - \rho z - \rho z^2 - \rho z^k]^2}$$

$$\therefore \ P'(1) = \frac{(1 - k\rho)\rho[1 + 2 + 3 + \cdots + k]}{(1 - k\rho)^2} = \frac{k(k + 1)\rho}{2(1 - k\rho)}$$

Average number of customers in the system $= \dfrac{k(k + 1)\rho}{2(1 - k\rho)} = L$, say

Average number of customers in the queue $= \displaystyle\sum_{n=1}^{\infty} (n - 1)p_n$

$$\therefore L_q = \sum_{n=1}^{\infty} n p_n - \sum_{n=1}^{\infty} p_n$$

$$= L - (1 - p_0) = L - k\rho$$

$$L = L_q + k\rho$$

Of course if $k = 1$ we have the result given just below (2.11).

6.2 Service consisting of k independent phases each having a negative exponential service with mean $1/\beta$

The mathematical model of the previous problem can be applied to the case of *single* random arrivals at rate α where each arrival needs service which is made up of k independent phases, each phase being negative exponential in type at rate β.

$p_n(t)$ denotes the probability that there are n service *phases* in the system at time t. Thus if $n = mk + j$ where $0 < j < k$, then this would correspond to $(m + 1)$ customers in the system with the customer being served still needing a further j phases of his service. Alternatively there are m customers in the queue and the customer being served still requires j phases of service (out of their k phases) to be completed.

As before (6.2) the steady-state equations for p_n (the probability that there are n service phases in the system) are:

$$-ap_0 + \beta p_1 = 0$$
$$-(\alpha + \beta)p_n + \beta p_{n+1} = 0 \qquad 1 \leqslant n \leqslant k - 1 \qquad (6.6)$$
$$ap_{n-k} - (\alpha + \beta)p_n + \beta p_{n+1} = 0 \qquad n \geqslant k$$

Thus again if $P(z) = \sum_{n=0}^{\infty} p_n z^n$ denotes the generating function for p_n

$$P(z) = \frac{\beta p_0 (1 - z)}{az^{k+1} - (\alpha + \beta)z + \beta} \qquad (6.7)$$

and since $P(1) = 1$, on using l'Hopital's rule for indeterminate forms, we obtain

$$P(1) = \underset{z \to 1}{\text{Limit}} \ \frac{-\beta p_0}{(k+1)az^k - (\alpha + \beta)} = \frac{\beta p_0}{\beta - ka} = 1$$

$$\therefore \ p_0 = \frac{\beta - ka}{\beta} = 1 - \frac{ka}{\beta} = 1 - k\rho \qquad (6.8)$$

Example 6.1

A quality control inspector has as his priority task the inspection of out-going production before it is passed on for further processing or returned for corrective action. The production arrives as a batch and these batches arrive at random at an average rate of 5/hour. On each batch he carries out three tests in order 1, 2, 3. The time taken for each test has a negative exponential distribution with mean 3 minutes.

In the steady-state situation p_n denotes the probability that there are n *tests* in the system. Show that the probability generating function for p_n has the form

$$P(z) = 1/(4 - z - z^2 - z^3)$$

Find

(a) the proportion of time that the inspector can devote to duties other than the inspection process;
(b) the probability that there is just one batch in the system undergoing its final test;
(c) the probability that there is just one batch in the system.

Here the tests correspond to the phases of the previous model and so $k = 3$. The arrival rate at 5/hour is 1/12 per minute ($\alpha = 1/12$) and the mean time for a test (phase) at 3 minutes ($= 1/\beta$), so that $\beta = 1/3$ (of a test per minute).

$$\therefore \ \rho = \frac{a}{\beta} = \left(\frac{1}{12}\right) \bigg/ \left(\frac{1}{3}\right) = \frac{1}{4}$$

$$\therefore \ p_0 = 1 - 3 \times \frac{1}{4} = \frac{1}{4}$$

Thus from (6.3) or (6.7),

$$P(z) = \frac{\dfrac{1}{4}(1-z)}{1 - \left(1 + \dfrac{1}{4}\right)z + \dfrac{1}{4}z^4} = \frac{1-z}{4 - 5z + z^4}$$

$$\therefore \; P(z) = \frac{(1-z)}{(1-z)(4 - z - z^2 - z^3)} = \frac{1}{4 - z - z^2 - z^3}$$

Note: we know the first denominator must have $(1-z)$ as a factor since $P(1)=1$. We observe that this is so.

To find p_n we need to expand $P(z)$ as a power series in z; p_n is the coefficient of z^n. In general this will be very difficult to find apart from a few small values of n.

$$P(z) = \frac{1}{4\left(1 - \dfrac{(z + z^2 + z^3)}{4}\right)}$$

$$= \frac{1}{4}\left\{1 + \left(\frac{z + z^2 + z^3}{4}\right) + \left(\frac{z + z^2 + z^3}{4}\right)^2 + \left(\frac{z + z^2 + z^3}{4}\right)^3 + \cdots\right\}$$

$$= \frac{1}{4}\left\{1 + z\left(\frac{1}{4}\right) + z^2\left(\frac{1}{4} + \frac{1}{16}\right) + z^3\left(\frac{1}{4} + \frac{2}{16} + \frac{1}{64}\right) + \cdots\right\}$$

Therefore, (a) we require p_0 and this is $1/4$ as we already know.
(b) We require p_1 and this is $1/4 \times 1/4 = 1/16$.
(c) We require $p_1 + p_2 + p_3$ and this is

$$\frac{1}{16} + \left(\frac{1}{16} + \frac{1}{64}\right) + \left(\frac{1}{16} + \frac{2}{64} + \frac{1}{256}\right)$$

$$= \frac{1}{256}\{16 + 16 + 4 + 16 + 8 + 1\} = \frac{61}{256}$$

The probability that there are n batches in the system is given by $p_{3n-2} + p_{3n-1} + p_{3n}$, a quantity which we can find in theory, although in practice it will be very difficult to do so. The power series expansion of $P(z)$ is not easy to say the least.

Example 6.2

Let us modify Example 6.1 so that the arrival rate $\alpha = 1/12$ still, but there are only two tests $(k=2)$ and the mean time for each test is 1 minute $(\beta = 1)$. Then $k=2$ and $\rho = \alpha/\beta = 1/12$. Then the probability generating function for p_n (probability 'n tests' in the system), is given by

$$P(z) = \frac{\left(1 - \frac{2}{12}\right)(1 - z)}{1 - \frac{13}{12}z + \frac{z^3}{12}}$$

(from (6.3) or (6.7))

$$= \frac{10(1 - z)}{12 - 13z + z^3} = \frac{10(1 - z)}{(1 - z)(12 - z - z^2)} = \frac{10}{12 - z - z^2}$$

$$= \frac{10}{12}\left(1 - \frac{(z + z^2)}{12}\right)^{-1}$$

$$= \frac{5}{6}\left[1 + \left(\frac{z + z^2}{12}\right) + \left(\frac{z + z^2}{12}\right)^2 + \left(\frac{z + z^2}{12}\right)^3 + \left(\frac{z + z^2}{12}\right)^4 + \cdots\right]$$

$$= \frac{5}{6}\left[1 + z\left(\frac{1}{12}\right) + z^2\left(\frac{1}{12} + \frac{1}{144}\right) + z^3\left(\frac{2}{144} + \frac{1}{1728}\right)\right.$$

$$\left. + z^4\left(\frac{1}{144} + \frac{3}{1728} + \frac{1}{20736}\right) + \cdots\right]$$

$$\therefore \ p_0 = \frac{5}{6}, \qquad p_1 = \frac{5}{72}, \qquad p_2 = \frac{65}{6 \times 144}, \qquad p_3 = \frac{5 \times 25}{6 \times 1728}, \qquad p_4 = \frac{5 \times 181}{6 \times 20736}, \text{etc.}$$

$\therefore$ Probability there is 1 batch in the system

$$= p_1 + p_2 = \frac{60 + 65}{6 \times 144} = \frac{125}{864}$$

$\therefore$ Probability there are 2 batches in the system

$$= p_3 + p_4 = \frac{5}{6} \cdot \frac{25 \times 12 + 181}{20736}$$

$$= \frac{5 \times 481}{6 \times 20736} = \frac{5 \times 481}{124416}$$

The etc. above is somewhat optimistic. The task gets quite complicated this way. But see Examples 13, question 2 for a hint.

The generating function can be used to find the mean number of tests in the system as $P'(z)|_{z=1}$.

$$P(z) = \frac{10}{12 - z - z^2}$$

[note $P(1) = 1$]

$$P'(z) = \frac{10(1 + 2z)}{(12 - z - z^2)^2}$$

$$\therefore \ P'(1) = \frac{10 \times 3}{10^2} = \frac{3}{10}$$

We can use the method of the imbedded process to find the generating function for π_n, the probability that a departing batch leaves n batches behind in the system.

If $Q(z) = \sum_{n=0}^{\infty} \pi_n z^n$ then we have the result (5.3) or (4.8)

$$Q(z) = \frac{(1-\rho)(1-z)\Phi(z)}{\Phi(z) - z}$$

However, we need to be careful. The ρ in the formula above is the arrival rate $\times$ mean service time of a *batch*

$$= \alpha \times \left(2 \times \frac{1}{\beta}\right) = \frac{1}{12} \times 2 \times 1 = \frac{1}{6} \; [k\rho \text{ as we defined } \rho \text{ in Example 6.2}]$$

$$\Phi(z) = B^*[\alpha(1-z)] \text{ from (4.9)}$$

where $B^*(s)$ is the Laplace transform of the service time of a batch.

Now the service time is the *sum* of the two times, one for each test. For one test the 'test time' T has p.d.f. $\beta e^{-\beta t}$. The Laplace transform of this, the p.d.f. of the test time for one phase, is

$$\int_0^{\infty} \beta e^{-\beta t} e^{-st} \, dt = \frac{\beta}{\beta + s} = \frac{1}{1 + \dfrac{s}{\beta}}$$

Then

$$B^*(s) = \left(\frac{1}{1 + \dfrac{s}{\beta}}\right)^2 \qquad \text{[convolution theorem]}$$

Then

$$\Phi(z) = 1 \bigg/ \left(1 + \frac{\alpha(1-z)}{\beta}\right)^2$$

In our case $\alpha = 1/12$, $\beta = 1$, so

$$\Phi(z) = 1 \bigg/ \left(1 + \frac{1}{12}(1-z)\right)^2$$

$$= 144/(13 - z)^2$$

$$\therefore \; Q(z) = \frac{5}{6}(1-z) \frac{144}{(13-z)^2} \bigg/ \left[\frac{144}{(13-z)^2} - z\right]$$

$$= \frac{120(1-z)}{144 - z(13-z)^2} = \frac{120(1-z)}{144 - 169z + 26z^2 - z^3} = \frac{120(1-z)}{(1-z)(144 - 25z + z^2)}$$

$$\therefore \; Q(z) = \frac{120}{144 - 25z + z^2} = \frac{120}{(z-9)(z-16)} = \sum_{n=0}^{\infty} \pi_n z^n$$

We note that $Q(1) = 120/(8 \times 15) = 1$ as it must be (just a check).

$$Q(0) = \frac{120}{9 \times 16} = \frac{10}{12} = \frac{5}{6} = \pi_0 \ (= p_0 \text{ of Example 6.2})$$

$$Q(z) = \frac{120}{7} \left[\frac{1}{z - 16} - \frac{1}{z - 9} \right] = \frac{120}{7} \left[\frac{1}{9 - z} - \frac{1}{16 - z} \right]$$

$$Q(z) = \frac{120}{7} \left[\frac{1}{9} \left(1 - \frac{z}{9} \right)^{-1} - \frac{1}{16} \left(1 - \frac{z}{16} \right)^{-1} \right]$$

$$= \frac{120}{7} \left[\frac{1}{9} \left\{ 1 + \frac{z}{9} + \frac{z^2}{9^2} + \frac{z^3}{9^3} + \cdots \right\} - \frac{1}{16} \left\{ 1 + \frac{z}{16} + \frac{z^2}{16^2} + \cdots \right\} \right]$$

$$\therefore \ \pi_n = \text{coeff. of } z^n = \frac{120}{7} \left[\left(\frac{1}{9} \right)^{n+1} - \left(\frac{1}{16} \right)^{n+1} \right]$$

$$\therefore \ \pi_0 = \frac{120}{7} \left(\frac{1}{9} - \frac{1}{16} \right) = \frac{120}{7} \cdot \frac{7}{144} = \frac{5}{6}$$

$$\pi_1 = \frac{120}{7} \left[\frac{1}{81} - \frac{1}{256} \right] = \frac{12 \times 10}{7} \times \frac{175}{9 \times 16 \times 9 \times 16}$$

$$= \frac{5 \times 25}{6 \times 144} = \frac{125}{864} \quad (p_1 + p_2 \text{ of Example 6.2})$$

$$\pi_2 = \frac{120}{7} \left[\frac{1}{729} - \frac{1}{4096} \right] = \frac{5 \times 24}{7} \times \frac{3367}{729 \times 4096}$$

$$= \frac{5 \times 481}{124416} \quad (= p_3 + p_4 \text{ of Example 6.2})$$

The mean number of batches (NOT tests) in the system is given by $Q'(1)$

$$Q'(z) = \frac{120}{7} \left[\frac{1}{(9 - z)^2} - \frac{1}{(16 - z)^2} \right]$$

$$\therefore \ Q'(1) = \frac{120}{7} \left(\frac{1}{64} - \frac{1}{225} \right) = \frac{120 \times 161}{7 \times 64 \times 225}$$

$$= \frac{120 \times 23}{64 \times 225} = \frac{23}{8 \times 15} = \frac{23}{120}$$

6.3 Examples 13

1 Customers in a queueing system are served in three sequential stages ($s = 1, 2, 3$), with the order of service being stage 3, stage 2, stage 1, for each customer. Each stage has a negative exponential service time distribution with mean 20 seconds and the stages are independent of one another. It is assumed that no time is lost between stages and only one customer is allowed in the service facility at any one time. Suppose customers arrive at random at a mean rate of one every 80 seconds, and $p_{n,s}$ represents the steady-state probability that there are n customers present and the customer in service is in stage s ($n = 1, 2, 3,...$). Show that the generating function defined as

$$G(y) = p_0 + \sum_{n=1}^{\infty} \sum_{s=1}^{3} y^{3(n-1)+s} p_{n,s}$$

where p_0 is the steady-state probability that there are no customers present, is given by $1/(4 - y - y^2 - y^3)$.

Show that on average there are 6 stages of service in the system and deduce that the mean time spent in the queue by a customer is 2 minutes Find the probability that there is:

(a) one customer in service in stage 3 and none waiting in the queue;
(b) one customer in service and none waiting in the queue.

2 For each article that arrives, a warehouse storeman has to undertake two jobs sequentially. When an article for storage arrives he first has to unpack it, and then he has to identify and store it. The time for each of these jobs has a negative exponential distribution with mean 2.5 minutes, and no time is lost between the two jobs. Articles arrive at random at a mean rate of 2/hour, and the storeman deals with them one at a time. Set up differential-difference equations relating the various states of the system. If steady-state conditions hold, and p_n denotes the probability that there are n jobs in the system at an arbitrary instant of time ($n = 0, 1, 2, ...$), show that the probability generating function, $P(z) = \sum_{n=0}^{\infty} z^n p_n$ is given by

$$P(z) = \frac{10}{(4 + z)(3 - z)}$$

and hence show that

$$p_n = \frac{10}{7} \left[\left(\frac{1}{3}\right)^{n+1} - \left(-\frac{1}{4}\right)^{n+1} \right]$$

Deduce the average number of jobs in the system. Find the probability that there is just one article in the system and this is being unpacked. If q_n denotes the probability that there are n *articles* in the system show that

$$q_n = p_{2n} + p_{2n-1} = \frac{120}{7} \left[\left(\frac{1}{9}\right)^{n+1} - \left(\frac{1}{16}\right)^{n+1} \right]$$

See the last part of Example 6.2 of this chapter.

6.4   Solutions to Examples 13

1 With the time unit as minutes, the arrival rate $\alpha = 3/4$ and the service rate (per stage) $\beta = 3$.

$p_{ns}(t) = \Pr(n$ customers present and customer in service in stage s at time $t)$ and $p_0(t) = \Pr($system is empty at time $t)$.

Then to first order in δt,

$$p_0(t + \delta t) = p_0(t)(1 - \tfrac{3}{4}\delta t) + p_{11}(t)3\delta t$$
$$p_{11}(t + \delta t) = p_{11}(t)(1 - \tfrac{3}{4}\delta t)(1 - 3\delta t) + p_{12}(t)3\delta t$$
$$p_{12}(t + \delta t) = p_{12}(t)(1 - \tfrac{3}{4}\delta t)(1 - 3\delta t) + p_{13}(t)3\delta t$$
$$p_{13}(t + \delta t) = p_0(t)\tfrac{3}{4}\delta t + p_{13}(t)(1 - \tfrac{3}{4}\delta t)(1 - 3\delta t) + p_{21}(t)3\delta t$$
$$p_{21}(t + \delta t) = p_{11}(t)\tfrac{3}{4}\delta t + p_{21}(t)(1 - \tfrac{3}{4}\delta t)(1 - 3\delta t) + p_{22}(t)3\delta t$$
$$p_{22}(t + \delta t) = p_{12}(t)\tfrac{3}{4}\delta t + p_{22}(t)(1 - \tfrac{3}{4}\delta t)(1 - 3\delta t) + p_{23}(t)3\delta t$$
$$p_{23}(t + \delta t) = p_{13}(t)\tfrac{3}{4}\delta t + p_{23}(t)(1 - \tfrac{3}{4}\delta t)(1 - 3\delta t) + p_{31}(t)3\delta t$$

etc.

Thus for the steady-state situation, after a little simplification

$$-p_0 + 4p_{11} = 0$$
$$-5p_{11} + 4p_{12} = 0$$
$$-5p_{12} + 4p_{13} = 0$$
$$p_0 - 5p_{13} + 4p_{21} = 0$$
$$p_{11} - 5p_{21} + 4p_{22} = 0$$
$$p_{12} - 5p_{22} + 4p_{23} = 0$$
$$p_{13} - 5p_{23} + 4p_{31} = 0$$

etc.

Thus multiplication of the above by y^0, y^1, y^2, y^3, etc. followed by addition gives

$$y^3 G(y) - 5[G(y) - p_0] + \frac{4[G(y) - p_0]}{y} - p_0 = 0$$

$$\therefore \; G(y)[y^4 - 5y + 4] = 4p_0(1 - y)$$

$$\therefore \; G(y) = \frac{4p_0}{(4 - y - y^2 - y^3)}$$

Since $G(1) = 1$, $4p_0 = 1$ so that finally we have

$$G(y) = \frac{1}{4 - y - y^2 - y^3}$$

$G'(y)\big|_{y=1}$ = expected number of *stages* in the system

$$= \frac{y + 2y + 3y^2}{(4 - y - y^2 - y^3)^2}\bigg|_{y=1} = 6$$

Thus on arriving, a customer finds on average 6 stages of service in front of them. Since each service stage or residual service stage (for an exponential distribution) takes on average 20 seconds, the average time spent in the queue is 2 minutes.

$$G(y) = \frac{1}{4}\left[1 - \left(\frac{y + y^2 + y^3}{4}\right)\right]^{-1}$$

$$= \frac{1}{4}\left[1 + \frac{1}{4}(y + y^2 + y^3) + \frac{1}{16}(y + y^2 + y^3)^2 + \frac{1}{64}(y + y^2 + y^3)^3 + \cdots\right]$$

$$= \frac{1}{4}\left[1 + y\left(\frac{1}{4}\right) + y^2\left(\frac{1}{4} + \frac{1}{16}\right) + y^3\left(\frac{1}{4} + \frac{2}{16} + \frac{1}{64}\right) + \cdots\right]$$

$$\therefore\ p_0 = \frac{1}{4}, \qquad p_{11} = \frac{1}{16}, \qquad p_{12} = \frac{5}{64}, \qquad p_{13} = \frac{16 + 8 + 1}{256} = \frac{25}{256}$$

Thus for (a) we require $p_{13} = 25/256$, and for (b) we require $p_{11} + p_{12} + p_{13} = 25/256 + 20/256 + 16/256 = 61/256$.

2 With the time unit as hours, the arrival rate $\alpha = 2$ and the service rate per job is $\beta = 24$. Thus with the same notation as in the previous question, we have to first order in δt

$$p_0(t + \delta t) = p_0(t)(1 - 2\delta t) + p_{11}(t)24\delta t$$
$$p_{11}(t + \delta t) = p_{11}(t)(1 - 2\delta t)(1 - 24\delta t) + p_{12}(t)24\delta t$$
$$p_{12}(t + \delta t) = p_0(t)2\delta t + p_{12}(t)(1 - 2\delta t)(1 - 24\delta t) + p_{21}(t)24\delta t$$
$$p_{21}(t + \delta t) = p_{11}(t)2\delta t + p_{21}(t)(1 - 2\delta t)(1 - 24\delta t) + p_{22}(t)24\delta t$$
etc.

Thus (with $'$ denoting differentiation with respect to t)

$$p_0'(t) = -2p_0(t) + 24p_{11}(t)$$
$$p_{11}'(t) = -26p_{11}(t) + 24p_{12}(t)$$
$$p_{12}'(t) = 2p_0(t) - 26p_{12}(t) + 24p_{21}(t)$$
$$p_{21}'(t) = 2p_{11}(t) - 26p_{21}(t) + 24p_{12}(t)$$
etc.

Thus in the steady-state we shall have

$$-2p_0 + 24p_{11} = 0$$
$$-26p_{11} + 24p_{12} = 0$$
$$2p_0 - 26p_{12} + 24p_{21} = 0$$
$$2p_{11} - 26p_{21} + 24p_{22} = 0$$
etc.

Define $P(z) = p_0 + \sum_{n=1}^{\infty}\sum_{s=1}^{2} z^{2(n-1)+s} p_{ns}\left(= \sum_{k=0}^{\infty} p_k z^k\right)$

Thus if we multiply the equations above by z^0, z^1, z^2, etc. and add, we obtain

$$2z^2 P(z) - 26[P(z) - p_0] + \frac{24[P(z) - p_0]}{z} - 2p_0 = 0$$

$$\therefore\ P(z)[2z^3 - 26z + 24] = 24p_0(1 - z)$$

$$\therefore\ P(z) = \frac{24p_0(1 - z)}{(2z^3 - 26z + 24)} = \frac{24p_0(1 - z)}{(1 - z)(24 - 2z - 2z^2)}$$

$$\therefore\ P(z) = \frac{12p_0}{12 - z - z^2}$$

Since $P(1) = 1$ we have $12p_0 = 10$ so that $p_0 = 5/6$.

$$\therefore\ P(z) = \frac{10}{12 - z - z^2} = \frac{10}{(4 + z)(3 - z)}$$

$$\therefore\ P(z) = \frac{10}{7}\left[\frac{1}{4 + z} + \frac{1}{3 - z}\right]$$

$$= \frac{10}{7}\left\{\frac{1}{4}\left(1 + \frac{z}{4}\right)^{-1} + \frac{1}{3}\left(1 - \frac{z}{3}\right)^{-1}\right\}$$

$$\therefore\ P(z) = \frac{10}{7}\left[\frac{1}{3}\left(1 + \frac{z}{3} + \left(\frac{z}{3}\right)^2 + \left(\frac{z}{3}\right)^3 + \cdots\right) + \frac{1}{4}\left(1 - \frac{z}{4} + \left(\frac{z}{4}\right)^2 - \left(\frac{z}{4}\right)^3 + \cdots\right)\right]$$

$$\therefore\ p_n = \frac{10}{7}\left[\left(\frac{1}{3}\right)^{n+1} - \left(-\frac{1}{4}\right)^{n+1}\right]$$

The average number of jobs in the system is given by the value of $\mathrm{d}P/\mathrm{d}z\,|_{z=1}$

$$\frac{\mathrm{d}P}{\mathrm{d}z} = \frac{10}{7}\left[\frac{1}{(3 - z)^2} - \frac{1}{(4 + z)^2}\right]$$

$$\therefore\ \frac{\mathrm{d}P}{\mathrm{d}z}\bigg|_{z=1} = \frac{10}{7}\left[\frac{1}{4} - \frac{1}{25}\right] = \frac{10}{7}\cdot\frac{21}{100} = \frac{3}{10}$$

The probability that there is an article in the system and that this article is being unpacked is given by p_{12}, i.e. the coefficient of z^2 in $P(z)$, i.e. p_2.

$$p_2 = \frac{10}{7}\left[\left(\frac{1}{3}\right)^3 + \left(\frac{1}{4}\right)^3\right] = \frac{10}{7}\left[\frac{1}{27} + \frac{1}{64}\right] = \frac{10 \times 91}{7 \times 27 \times 64}$$

$$= \frac{5 \times 13}{27 \times 32} = \frac{65}{864}$$

For there to be n articles in the system, the nth article can either be being unpacked, or identified and stored. Thus

$$q_n = p_{2n-1} + p_{2n} \qquad (n \geq 1)$$

$$\therefore q_n = \frac{10}{7} \cdot \frac{1}{3} \left[\left(\frac{1}{3}\right)^{2n-1} + \left(\frac{1}{3}\right)^{2n} \right] + \frac{10}{7} \cdot \frac{1}{4} \left[\left(\frac{1}{4}\right)^{2n} - \left(\frac{1}{4}\right)^{2n-1} \right]$$

$$= \frac{10}{7} \cdot \frac{3}{3} \left(\frac{1}{9}\right)^{n} + \frac{10}{7} \cdot \frac{1}{3} \left(\frac{1}{9}\right)^{n} + \frac{10}{7} \cdot \frac{1}{4} \left(\frac{1}{16}\right)^{n} - \frac{10}{7} \cdot \frac{4}{4} \left(\frac{1}{16}\right)^{n}$$

$$= \frac{10}{7} \cdot \frac{4}{3} \left(\frac{1}{9}\right)^{n} - \frac{10}{7} \cdot \frac{3}{4} \left(\frac{1}{16}\right)^{n}$$

$$\therefore q_n = \frac{120}{7} \left[\left(\frac{1}{9}\right)^{n+1} - \left(\frac{1}{16}\right)^{n+1} \right]$$

7
Random arrivals, block service, constant interval between the times at which service is available

We consider a system in which customers arrive at random at an average rate λ, say, and wait for service. Up to N customers can be served at a time when service is available and this is so at constant intervals of c units of time. Examples might be a cable-car lift system where the cars arrive at, say, 1-minute intervals and can accommodate three customers ($N = 3$, $c = 1$). A 'Paternoster' lift is another example (Fig. 7.1). This consists of a number of platforms on a continuous belt and passengers just step onto the platform as it passes a floor. The empty platforms arrive at, say, 15-second intervals (c) at the bottom floor, and may be able to carry one or even two customers ($N = 1$ or perhaps $N = 2$).

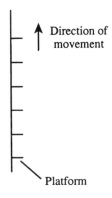

Direction of movement

Platform

Fig. 7.1

A different viewpoint is that we have random arrivals at rate λ, N servers who all work together and constant service time c for each service by each server. Other examples of continuous conveyor systems whose carriages arrive at constant intervals (c) and which can carry up to N items at a time, can be modelled in this way.

The arrival process is a Poisson stream with average arrival rate λ. We consider the steady-state situation (which is assumed to exist) and let p_n ($n = 0, 1, 2, \ldots$) denote the probability that n customers are waiting in the queue at the moment service becomes available. Then if we consider two consecutive such instants separated by c time units

$$p_0 = (p_0 + p_1 + \cdots + p_N)e^{-\lambda c} \qquad (7.1)$$

There are none waiting at the start of service if at the previous start of service there were no more than N waiting (and of course they were all served) *and* no further customers arrive in the interval between service being available.

$$p_1 = (p_0 + p_1 + \cdots + p_N)\frac{\lambda c e^{-\lambda c}}{1!} + p_{N+1}e^{-\lambda c} \qquad (7.2)$$

There is one waiting at the start of service if at the start of the previous service there were no more than N waiting *and* one customer arrived in the interval until the next service *or* there were exactly $N+1$ customers waiting at the start of the previous service (one was still left after this service) and no customers arrived in the interval until the start of the next service. Similarly

$$p_2 = (p_0 + p_1 + \cdots + p_N)\frac{(\lambda c)^2 e^{-\lambda c}}{2!} + p_{N+1}\frac{\lambda c e^{-\lambda c}}{1!} + p_{N+2}e^{-\lambda c}$$

Here there are three mutually exclusive alternatives:

(i) up to N waiting at the previous service *and* two arrivals in interval, *or*
(ii) $N+1$ waiting at the previous service *and* one arrival in interval, *or*
(iii) $N+2$ waiting at the previous service *and* no arrivals in interval.

Thus the system of equations for p_n is

$$p_0 = (p_0 + p_1 + \cdots + p_N)e^{-\lambda c}$$

$$p_1 = (p_0 + p_1 + \cdots + p_N)\frac{\lambda c e^{-\lambda c}}{1!} + p_{N+1}e^{-\lambda c}$$

$$p_2 = (p_0 + p_1 + \cdots + p_N)\frac{(\lambda c)^2 e^{-\lambda c}}{2!} + p_{N+1}\frac{\lambda c e^{-\lambda c}}{1!} + p_{N+2}e^{-\lambda c} \qquad (7.3)$$

$$p_n = (p_0 + p_1 + \cdots + p_N)\frac{(\lambda c)^n e^{-\lambda c}}{n!} + p_{N+1}\frac{(\lambda c)^{n-1} e^{-\lambda c}}{(n-1)!} + \cdots + p_{N+n}e^{-\lambda c}$$

etc.

We define the generating function

$$P(z) = \sum_{n=0}^{\infty} p_n z^n \qquad (7.4)$$

If we multiply the above equations by z^0, z^1, z^2, ... etc. and add we obtain

$$\sum_{n=0}^{\infty} p_n z^n = (p_0 + p_1 + \cdots + p_N)e^{-\lambda c}\sum_{n=0}^{\infty}\frac{(\lambda c z)^n}{n!} + e^{-\lambda c}p_{N+1}z\sum_{r=0}^{\infty}\frac{(\lambda c z)^r}{r!}$$

$$+ e^{-\lambda c}p_{N+2}z^2\sum_{r=0}^{\infty}\frac{(\lambda c z)^r}{r!} + \cdots$$

$$\therefore \ P(z) = (p_0 + p_1 + \cdots + p_N)e^{-\lambda c}e^{\lambda cz} + p_{N+1}ze^{-\lambda c}e^{\lambda cz} + p_{N+2}z^2e^{-\lambda c(1-z)}$$

$$+ \cdots + p_{N+k}z^k e^{-\lambda c(1-z)} + \cdots$$

$$\therefore \ P(z) = (p_0 + p_1 + \cdots + p_N)e^{-\lambda c(1-z)}$$

$$+ e^{-\lambda c(1-z)}\left[\frac{P(z) - p_0 - p_1 z - p_2 z^2 - \cdots - p_N z^N}{z^N}\right]$$

$$\therefore \ P(z)e^{\lambda c(1-z)} = (p_0 + p_1 + \cdots + p_N) + \frac{[P(z) - p_0 - p_1 z - \cdots - p_N z^N]}{z^N}$$

$$\therefore \ P(z)[z^N e^{\lambda c(1-z)} - 1] = (p_0 + p_1 + \cdots + p_N)z^N - p_0 - p_1 z - p_2 z^2 - \cdots - p^N z^N$$

$$\therefore \ P(z) = \frac{(p_0 + p_1 + \cdots + p_N)z^N - p_0 - p_1 z - p_2 z^2 - \cdots - p_N z^N}{z^N e^{\lambda c(1-z)} - 1} \tag{7.5}$$

We have an expression for $P(z)$ but it is one that involves many of the probabilities (namely, p_0, p_1, ..., p_N) which are constituents of $P(z)$ itself. However, we have some knowledge of $P(z)$ and we can use this to advantage.

$$P(z) = p_0 + p_1 z + p_2 z^2 + \cdots + p_n z^n + \cdots$$

where for each n, $0 \le p_n \le 1$ and $\sum_{n=0}^{\infty} p_n = 1$.

Thus $P(z)$ is a well-behaved, convergent and hence bounded function for all values of z (real or complex) within the unit circle $|z| \le 1$. We have in the past used the condition that $P(1) = 1$ to obtain an expression for p_0 (see (2.7), (5.3) and (6.4) for example).

Now it follows that if the denominator has any zeros inside or on the unit circle then the numerator must have the same zeros. Of course we see at once that $z = 1$ is a zero of the denominator $z^N \exp\{\lambda c(1 - z)\} - 1$. This is certainly zero when $z = 1$. It can also be seen that the numerator

$$(p_0 + p_1 + \cdots + p_N)z^N - p_0 - p_1 z - p_2 z - \cdots - p_N z^N$$

is zero when $z = 1$. This has to be so since $P(1) = 1$. The numerator is a polynomial of degree N in z and so will have N zeros (roots). One of these is $z = 1$ as we have just seen. The others z_1, z_2, ..., z_{N-1} all lie within the unit circle and so have modulus less than 1. (These must coincide with the roots of the denominator which lie inside the unit circle.) We can see that these roots are all within the unit circle, for if we assume $|z_i| = \gamma > 1$, then since z_i is a zero of the numerator,

$$(p_0 + p_1 + \cdots + p_N)z^N = p_0 + p_1 z + p_2 z^2 + \cdots + p_N z^N$$

$$\therefore \ |(p_0 + p_1 + \cdots + p_N)z^N| = |p_0 + p_1 z + p_2 z^2 + \cdots + p_N z^N|$$

Now on the L.H.S. we have $(p_0 + p_1 + \cdots + p_N)\gamma^N$ and for the R.H.S.

$$|p_0 + p_1 z + p_2 z^2 + \cdots + p_N z^N| \le |p_0| + |p_1 z| + |p_2 z^2| + \cdots + |p_N z^N|$$

$$\le p_0 + p_1 \gamma + p_2 \gamma^2 + \cdots + p_N \gamma^N$$

$$< p_0 \gamma^N + p_1 \gamma^N + p_2 \gamma^N + p_N \gamma^N \qquad \text{since } \gamma > 1$$

Thus

$$(p_0 + p_1 + \cdots + p_N)\gamma^N < (p_0 + p_1 + \cdots + p_N)\gamma^N$$

which is impossible. Thus the numerator is zero at the N values $1, z_1, z_2, \ldots, z_{N-1}$ within and on the unit circle. The denominator, $z^N \exp\{\lambda c(1-z)\} - 1$, is zero for these same values and of course it is by solving this equation, which involves N, λ and c, that we can in principle find these roots. We can use Rouche's theorem, from Complex Variable Theory, to show that the equation $z^N \exp\{\lambda c(1-z)\} - 1$ has N zeros within and on the unit circle provided $\lambda c < N$. This condition is the one that we should expect in order for a steady-state solution to exist. The mean number of new arrivals in time c is less than N, the maximum number who can be served in a block. The numerator being a polynomial of degree N can be written as

$$K(z-1)(z-z_1)(z-z_2) \ldots (z-z_{N-1})$$

where K is a constant. We can find K since $P(1) = 1$

$$\therefore \text{ Limit}_{z \to 1} \frac{K(z-1)(z-z_1)(z-z_2) \ldots (z-z_{N-1})}{z^N e^{\lambda c(1-z)} - 1} = 1$$

This gives an indeterminate form $0/0$ so we must use l'Hopital's rule (again). The derivative of the denominator is

$$N z^{N-1} e^{\lambda c(1-z)} - \lambda c z^N e^{\lambda c(1-z)}$$

The derivative of the numerator can be written as

$$K(z-z_1)(z-z_2) \ldots (z-z_{N-1}) + (z-1) \times \text{another polynomial}$$

This follows from using the rule for differentiating a product. Thus on substituting $z = 1$ we obtain

$$K(1-z_1)(1-z_2) \ldots (1-z_{N-1}) = N - \lambda c$$

Thus finally

$$P(z) = \frac{(N-\lambda c)(z-1)(z-z_1) \ldots (z-z_{N-1})}{(1-z_1) \ldots (1-z_{N-1})} \cdot \frac{1}{z^N e^{\lambda c(1-z)} - 1} \tag{7.6}$$

Of course $z_1, z_2, \ldots, z_{N-1}$ are the roots of $z^N \exp\{\lambda c(1-z)\} - 1 = 0$ which lie within the unit circle. We can find these, and since these roots coincide with the roots of the numerator, we can use this information in principle to find $p_0, p_1, \ldots, p_{N-1}$.

7.1 The proportion of customers who don't have to wait beyond the first time that service becomes available

$$\Pr(W = 0) = \Pr(\text{zero delay}) = \sum_{i=0}^{N-1} p_i = a_{N-1} \text{ say.}$$

Let

$$Q(z) = \sum_{k=0}^{\infty} a_k z^k \text{ where } a_k = \sum_{i=0}^{k} p_i$$

Then since $a_k - a_{k-1} = p_k$

$$Q(z)(1-z) = P(z)$$

$$\therefore Q(z) = \frac{P(z)}{1-z}$$

Then a_{N-1} will be the coefficient of z^{N-1} in $Q(z)$. Now from our expression for $P(z)$, namely (7.6)

$$Q(z) = \frac{(N - \lambda c)(z - z_1)(z - z_2) \dots (z - z_{N-1})}{(1 - z_1)(1 - z_2) \dots (1 - z_{N-1})} \cdot \frac{1}{1 - z^N e^{\lambda c (1 - z)}}$$

The coefficient of z^{N-1} in this expression is

$$\frac{(N - \lambda c)}{(1 - z_1)(1 - z_2) \dots (1 - z_{N-1})} = a_{N-1}$$

This result follows since the coefficient of z^{N-1} in $(z - z_1)(z - z_2) \dots (z - z_{N-1})$ is 1 and the constant term in $1/[1 - z^N \exp\{\lambda c(1 - z)\}]$ is also 1. The probability that a customer does not have to wait beyond the time that the service facility first becomes available is

$$\frac{(N - \lambda c)}{(1 - z_1)(1 - z_2) \dots (1 - z_{N-1})} \tag{7.7}$$

which is easily found if $z_1, z_2, \dots, z_{N-1}$ are known.

7.2 The special case when N = 1

Example 7.1

Finished articles from the production line in a factory are carried to the despatch section on an 'endless' conveyor system. Empty compartments on this conveyor belt arrive at the end of the production line at constant intervals of c minutes. Each compartment can hold just one article. If there are no articles waiting to be carried when a compartment arrives, then that compartment continues in the empty state to the despatch section. If there are one or more articles waiting when a compartment arrives, then just one is loaded into the compartment. Finished articles to be conveyed arrive at the end of the production line at random at an average rate of λ per minute, where $\lambda c < 1$.

  Suppose p_n denotes the probability that n articles are waiting at the end of the production line when an empty compartment arrives. Present a clear argument to justify the equations:

$$p_0 = (p_0 + p_1)e^{-\lambda c}$$

$$p_1 = (p_0 + p_1)\lambda c e^{-\lambda c} + p_2 e^{-\lambda c}$$

$$p_2 = (p_0 + p_1)\frac{(\lambda c)^2}{2!} e^{-\lambda c} + p_2 \lambda c e^{-\lambda c} + p_3 e^{-\lambda c}$$

$$p_n = (p_0 + p_1)\frac{(\lambda c)^n e^{-\lambda c}}{n!} + p_2 \frac{(\lambda c)^{n-1} e^{-\lambda c}}{(n - 1)!} + p_3 \frac{(\lambda c)^{n-2} e^{-\lambda c}}{(n - 2)!}$$

$$+ \dots + p_n \lambda c e^{-\lambda c} + p_{n+1} e^{-\lambda c}$$

etc.

Hence show that if

$$P(z) = \sum_{n=0}^{\infty} p_n z^n$$

then

$$P(z) = \frac{p_0(1-z)}{1 - ze^{\lambda c(1-z)}}$$

If $\lambda = 1$ and $c = 1/2$, find the proportion of compartments which are empty on arrival at despatch.

Find also the mean number of articles waiting when a compartment arrives at the end of the production line.

For the solution we have

$$p_0 = (p_0 + p_1)e^{-\lambda c}$$

There are none waiting when a compartment arrives if there were no more than one waiting when the previous compartment arrived (so none were left behind) and no articles arrived during the interval between the arrival of the two compartments

$$p_1 = (p_0 + p_1)\lambda ce^{-\lambda c} + p_2e^{-\lambda c}$$

One is waiting when a compartment arrives if there were no more than one waiting when the previous compartment arrived and one article arrived in the interval or two were waiting when the previous compartment arrived (so one got left behind) and no further articles arrived in the interval.

This pattern is reflected in the form of the remaining equations. Those left behind plus the new arrivals make up the total found by the next compartment. Thus we have

$$p_0 = (p_0 + p_1)e^{-\lambda c}$$

$$p_1 = (p_0 + p_1)\lambda ce^{-\lambda c} + p_2e^{-\lambda c}$$

$$p_2 = (p_0 + p_1)\frac{(\lambda c)^2}{2!}e^{-\lambda c} + p_2\lambda ce^{-\lambda c} + p_3e^{-\lambda c}$$

etc.

We multiply these equations in turn by z^0, z^1, z^2, etc. and add so that with

$$P(z) = \sum_{n=0}^{\infty} p_n z^n$$

$$P(z) = (p_0 + p_1)e^{-\lambda c}e^{\lambda cz} + p_2ze^{-\lambda c}e^{\lambda cz} + p_3z^2e^{-\lambda c}e^{\lambda cz} + \cdots$$

column 1   column 2   column 3   etc.

$$\therefore\ P(z)e^{\lambda c(1-z)} = (p_0 + p_1) + p_2z + p_3z^2 + p_4z^3 + \cdots$$

$$= (p_0 + p_1) + \frac{[P(z) - (p_0 + p_1z)]}{z}$$

$$\therefore\ P(z)ze^{\lambda c(1-z)} = (p_0 + p_1)z + P(z) - p_0 - p_1z$$

$$\therefore\ P(z) = \frac{p_0(z-1)}{ze^{\lambda c(1-z)} - 1}$$

Both numerator and denominator are zero when $z = 1$, so to use the condition $P(1) = 1$ we have to use l'Hopital's rule

$$P(1) = \underset{z \to 1}{\text{Limit}} \ \frac{p_0}{e^{\lambda c(1-z)} - \lambda c z e^{\lambda c(1-z)}} = \frac{p_0}{1 - \lambda c} = 1$$

$$\therefore \ p_0 = 1 - \lambda c$$

Thus

$$P(z) = \frac{(1 - \lambda c)(z - 1)}{z e^{\lambda c(1-z)} - 1}$$

In our case $\lambda = 1$ and $c = 1/2$ so that $p_0 = 1 - 1/2 = 1/2$ and this is the proportion of compartments that leave empty. To find the expected number of articles waiting when a compartment arrives we need to evaluate $P'(1)$.

$$P(z) = \frac{\dfrac{1}{2}(z - 1)}{z e^{(1-z)/2} - 1}$$

$$P'(z) = \frac{1}{2} \cdot \frac{\left\{ (z e^{(1-z)/2} - 1) - (z - 1)\left[e^{(1-z)/2} - \dfrac{1}{2} z e^{(1-z)/2} \right] \right\}}{(z e^{(1-z)/2} - 1)(z e^{(1-z)/2} - 1)}$$

$P'(1)$ takes the form $0/0$, so we need to use l'Hopital's rule.

$$P'(1) =$$

$$\frac{1}{2} \underset{z \to 1}{\text{Limit}} \ \frac{\left[e^{(1-z)/2} - \dfrac{1}{2} z e^{(1-z)/2} \right] - \left[e^{(1-z)/2} - \dfrac{1}{2} z e^{(1-z)/2} \right] - (z - 1)\left[e^{(1-z)/2} + \dfrac{1}{4} z e^{(1-z)/2} \right]}{2(z e^{(1-z)/2} - 1)\left(e^{1/2(1-z)} - \dfrac{1}{2} z e^{1(1-z)/2} \right)}$$

$$= \frac{1}{2} \underset{z \to 1}{\text{Limit}} \ \frac{z - 1}{(z e^{(1-z)/2} - 1)} \cdot \frac{\left[e^{(1-z)/2} - \dfrac{1}{4} z e^{(1-z)/2} \right]}{\left(e^{(1-z)/2} - \dfrac{1}{2} z e^{(1-z)/2} \right)}$$

$$= \frac{1}{2} \times 1 \times \frac{\dfrac{3}{4}}{\dfrac{1}{2}} = \frac{3}{4}$$

where we have used the result that $P(1) = 1$ in resolving the first term of the limit.

A simpler alternative is to expand $P(z)$ as a power series in y $(=z-1)$ when $P'(1)$ is the coefficient of y. [Refer back to Chapter 1 on generating functions.] $P(z)$ as a function of y

$$= \frac{\frac{1}{2}y}{(1+y)e^{-y/2} - 1} = \frac{\frac{1}{2}y}{(1+y)\left(1 - \frac{1}{2}y + \frac{1}{8}y^2 - \cdots\right) - 1}$$

$$= \frac{\frac{1}{2}y}{1 + \frac{1}{2}y - \frac{3}{8}y^2 + \cdots - 1} = \frac{\frac{1}{2}y}{\frac{1}{2}y\left(1 - \frac{3}{4}y + \cdots\right)}$$

$$= 1 + \frac{3}{4}y + \cdots$$

Thus the coefficient of y is $3/4$ and this is the same as $P'(1)$ and gives the mean number of articles waiting when a compartment arrives.

The algebra above is in both cases a repeat of the working for question 4 of Examples 2 in the case where $\rho = 1/2$.

Example 7.2

Figure 7.2 shows a road junction which is controlled by traffic lights. The lights allow traffic from the minor road to turn right onto the major road in safety. Vehicles on the minor road arrive at the junction at random at an average rate of one every 2 minutes. The lights are timed to change to green at 2-minute intervals and remain green for long enough to allow two vehicles to join the major road. Find

  (i) the average number of vehicles waiting at the lights when they turn to green;
 (ii) the proportion of vehicles who can join the main road at the first green light following their arrival.

We shall use the notation and results which we have developed in the current section so that: $p_n \equiv \Pr(n$ vehicles are waiting at the lights when they turn green), $\lambda = 1/2$ (vehicle/ minute), $c = 2$ and $N = 2$. Thus from (7.6) the generating function for the p_n is

$$P(z) = \frac{(2-1)(z-1)(z-z_1)}{1 - z_1} \cdot \frac{1}{z^2 e^{(1-z)} - 1}$$

where z_1 is the root of $z^2 \exp\{(1-z)\} - 1$ which lies within the unit circle.

It is not difficult to verify that $z_1 = -0.477$. This result can be obtained using a suitable numerical procedure or a suitable computer algebra package such as DERIVE or MAPLE.

To find the mean number in the system we need to evaluate $P'(z)|_{z=1}$. It is easy to see that this will result in a complicated $0/0$ indeterminate form. A simpler (though not trivial) procedure is to use the method explained in Chapter 1, and in particular the solution to question 4 of Examples 2.

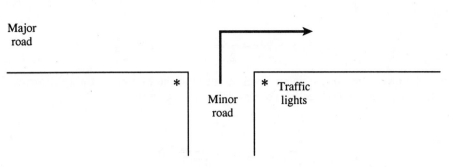

Fig. 7.2

Thus we expand $P(z)$ as a power series in $y = z - 1$ and find the coefficient of y.

$$P(z) = \frac{(z-1)(z-z_1)}{1-z_1} \frac{1}{z^2 e^{(1-z)} - 1}$$

$$= \frac{y(1 - z_1 + y)}{(1 - z_1)[(1 + y)^2 e^{-y} - 1]}$$

since $z = y + 1$. Thus $P(z)$ (as a function of y)

$$= \frac{y\left(1 + \dfrac{y}{1 - z_1}\right)}{\left\{(1 + 2y + y^2)\left(1 - y + \dfrac{y^2}{2} - \dfrac{y^3}{6} + \cdots\right) - 1\right\}}$$

$$= \frac{y\left(1 + \dfrac{y}{1 - z_1}\right)}{1 + 2y - y + y^2\left(1 - 2 + \dfrac{1}{2}\right) + \cdots - 1}$$

$$= \frac{y\left(1 + \dfrac{y}{1 - z_1}\right)}{y\left(1 - \dfrac{y}{2} + \cdots\right)}$$

$$= \left(1 + \frac{y}{1 - z_1}\right)\left(1 + \frac{y}{2} + \cdots\right)$$

$$= 1 + y\left(\frac{1}{1 - z_1} + \frac{1}{2}\right) + \cdots$$

and fortunately we do not need to consider higher powers of y. Thus the mean number of vehicles waiting is

$$\frac{1}{2} + \frac{1}{1 - z_1} \simeq \frac{1}{2} + \frac{1}{1.477} = 1.18$$

To find the proportion of vehicles which proceed at the first green light following their arrival, we can use (7.7) directly. The required proportion is

$$\frac{\left(2 - 2 \times \dfrac{1}{2}\right)}{1 + 0.477} = \frac{1}{1.477} \simeq 0.677$$

7.3   Examples 14

1   Barges arrive at random at a lock on a canal at mean rate λ in each direction. The lock can hold up to a maximum of N barges at a time. If there are no barges waiting from one direction when the barges from the other direction leave the lock, the lock-keeper still continues with the cycle of operations. The time taken from entry to the lock of barges from one direction to the entry of barges from the same direction has constant value c. If p_n is the probability that there are n barges waiting to enter the lock from one direction just before the cycle of operations begins from that direction, and $G(z) = \sum_{n=0}^{\infty} z^n p_n$, show that

$$G(z) = \frac{N(1 - \rho)(1 - z)}{1 - z^N e^{N\rho(1 - z)}} \prod_{i=1}^{N-1} \frac{z - z_i}{1 - z_i} \qquad \left(\rho = \frac{\lambda c}{N}\right)$$

where z_i $(i = 1, 2, \ldots, N - 1)$ are the roots of $1 - z^N \exp\{N\rho(1 - z)\} = 0$ which lie within the unit circle. (You may assume that this equation has exactly N roots within and on the unit circle, including the root at $z = 1$, provided that $\rho < 1$.)

If $\lambda = 6/\text{hour}$, $c = 10$ minutes, and $N = 2$, show that the mean number of barges waiting from one direction just before the start of a cycle of lock operations from that direction is $(3 - z_1)/2(1 - z_1)$, and verify that the value of z_1 is approximately -0.48.

2   You should assume that the equation $z^N = \exp\{N\rho(z - 1)\}$, ($\rho$ real) where N is a positive integer, has exactly N roots within and on the unit circle, provided that $\rho < 1$.

An aerial chair ride at a fairground operates as follows. Chairs which can hold two passengers arrive at the starting point empty at constant time intervals of 30 seconds. If there are two or more passengers waiting when a chair arrives then the first two passengers in the waiting line get in the chair as it passes. If there is just one passenger waiting then that person gets in the chair. If there are no passengers waiting the chair continues on the ride empty. Passengers for the chair ride arrive at the starting point at random at an average rate of three per minute.

If p_n is the probability that n passengers are waiting just before a chair arrives and we define the generating function, $P(z) = \sum_{k=0}^{\infty} z^k p_k$, show that

$$P(z) = \frac{(z - 1)(z - z_1)}{2(1 - z_1)\{z^2 e^{3(1 - z)/2} - 1\}}$$

where z_1 is the real root of $z^2 = \exp\{3(z-1)/2\}$ which lies within the unit circle. Verify that $z_1 = -0.360$ to 3 decimal places, and hence find the proportion of empty chairs on the ride.

Find the mean number of passengers who are waiting at the start of the ride just before a chair arrives.

7.4 Solutions to Examples 14

1 The probability that j barges arrive during an interval of duration c is given by

$$\frac{(\lambda c)^j}{j!} e^{-\lambda c}$$

Thus the equations for p_n are (with $a_N = p_0 + p_1 + \cdots + p_N$)

$$p_0 = a_N e^{-\lambda c}$$

$$p_1 = a_N \frac{\lambda c e^{-\lambda c}}{1!} + p_{N+1} e^{-\lambda c}$$

$$p_2 = a_N \frac{(\lambda c)^2 e^{-\lambda c}}{2!} + p_{N+1} \frac{\lambda c e^{-\lambda c}}{1!} + p_{N+2} e^{-\lambda c}$$

$$p_j = a_N \frac{(\lambda c)^j e^{-\lambda c}}{j!} + p_{N+1} \frac{(\lambda c)^{j-1} e^{-\lambda c}}{(j-1)!} + \cdots + p_{N+j} e^{-\lambda c}$$

etc.

Thus if we multiply the above equations by $z^0, z^1, z^2, \ldots, z^j, \ldots$ and add

$$G(z) = a_N e^{-\lambda c(1-z)} + \left[\frac{G(z) - (p_0 + p_1 z + p_2 z^2 + \cdots + p_N z^N)}{z^N} \right] e^{-\lambda c(1-z)}$$

$$\therefore \; z^N G(z) = a_N z^N e^{-N\rho(1-z)} + G(z) e^{-N\rho(1-z)} - (p_0 + p_1 z + \cdots + p_N z^N) e^{-N\rho(1-z)}$$

$$\therefore \; G(z)[1 - z^N e^{N\rho(1-z)}] = p_0 + p_1 z + \cdots + p_N z^N - z^N(p_0 + p_1 + \cdots + p_N)$$

$$\therefore \; G(z) = \frac{p_0 + p_1 z + p_2 z^2 + \cdots + p_N z^N - z^N(p_0 + p_1 + \cdots + p_N)}{1 - z^N e^{N\rho(1-z)}}$$

The zeros of the denominator must coincide with those of the numerator. One zero of the denominator is $z = 1$ and this is clearly a zero of the numerator. The others which arise within the unit circle are at $z_1, z_2, \ldots, z_{N-1}$. We can therefore assume that the numerator takes the form $\kappa(z-1)(z-z_1)(z-z_2) \ldots (z-z_{N-1})$. Since $\text{Limit}_{z \to 1} \, G(z) = 1$

$$\text{Limit}_{z \to 1} \; \frac{\kappa(z-1)(z-z_1) \ldots (z-z_{N-1})}{1 - z^N e^{N\rho(1-z)}} = 1$$

and using l'Hopital's rule this gives

$$\frac{\kappa(1-z_1)(1-z_2) \ldots (1-z_{N-1})}{-N(1-\rho)} = 1$$

so that

$$\kappa = \frac{-N(1-\rho)}{(1-z_1)(1-z_2)\dots(1-z_{N-1})}$$

$$\therefore\ G(z) = \frac{N(1-\rho)(1-z)\prod\limits_{i=1}^{N-1}\left(\dfrac{z-z_i}{1-z_i}\right)}{1-z^N e^{N\rho(1-z)}}$$

With time measured in minutes, $\lambda = 1/10$, $c = 10$, $N = 2$, so that $\rho = 1/2$, $N\rho = 1$ and $N(1-\rho) = 1$.

$$\therefore\ G(z) = \frac{(z-1)(z-z_1)}{(1-z_1)[z^2 e^{(1-z)} - 1]}$$

When $z = -0.48$, $z^2\exp\{(1-z)\} - 1 = 0.2304 \times 4.392945681 - 1 \approx 0.01213$. Thus $z_1 = -0.480$ is the root required.

We can find the mean number of waiting barges as the value of $G'(z)|_{z=1}$. However, when we do this we will find that $G'(1)$ takes the form $0/0$ and two applications of l'Hopital's rule are required before the given result emerges. An easier procedure is to find the coefficient of y ($=z-1$) in the power series expansion of $G(z)$ as a power series in $(z-1)$ (refer to Chapter 1 and Examples 2, question 4 and to the last example), but we still have to take care with the algebra.

$$G(z) = \frac{(z-1)(z-z_1)}{(z^2 e^{1-z} - 1)(1-z_1)}$$

with $z - 1 = y$, $z = y + 1$.

$$\therefore\text{ As a function of } y,\ G(z) = \frac{y(1 - z_1 + y)}{(1-z_1)[(1+y)^2 e^{-y} - 1]}$$

$$= \frac{y\left(1 + \dfrac{y}{1-z_1}\right)}{\left\{(1+2y+y^2)\left(1 - y + \dfrac{y^2}{2} - \dfrac{y^3}{6} + \cdots\right) - 1\right\}}$$

$$= \frac{y\left(1 + \dfrac{y}{1-z_1}\right)}{1 + 2y - y + y^2\left(1 - 2 + \dfrac{1}{2}\right) + \cdots - 1}$$

$$= \frac{y\left(1 + \dfrac{y}{1-z_1}\right)}{y\left(1 - \dfrac{y}{2} + \cdots\right)}$$

This is

$$\frac{1 + \dfrac{y}{1 - z_1}}{1 - \dfrac{y}{2} + \cdots} = \left(1 + \frac{y}{1 - z_1}\right)\left(1 + \frac{y}{2} + \left(\frac{y}{2}\right)^2 + \cdots\right)$$

The coefficient of y is

$$\frac{1}{1 - z_1} + \frac{1}{2} = \frac{3 - z_1}{2(1 - z_1)}$$

and this is the mean value required.

2 We can model this problem in precisely the same way as question 1 up to the point where we obtain

$$P(z) = \frac{N(1 - \rho)(1 - z)\displaystyle\prod_{i=1}^{N-1}\left(\frac{z - z_i}{1 - z_i}\right)}{1 - z^N e^{N\rho(1 - z)}}$$

Now, with the time unit as minutes, $\lambda = 3$, $c = 1/2$, $N = 2$, so that $\rho = 3/4$, $1 - \rho = 1/4$ and $N\rho = 3/2$. Thus

$$P(z) = \frac{(z - 1)(z - z_1)}{2(1 - z_1)\{z^2 e^{3(1 - z)/2} - 1\}}$$

where z_1 is the real root of $z^2 \exp\{3(1 - z)/2\} = 1$ which lies within the unit circle. When $z = -0.360$, $z^2 \exp\{3(1 - z)/2\} = 0.1296 \times e^{2.04} = 0.9967$. Thus $z_1 = -0.360$ is the required root.

$$p_0 = P(0) = \frac{z_1}{2(-1) \times 1.36} = \frac{-z_1}{2.72} = 0.1324$$

and this gives the proportion of empty chairs on the ride. The evaluation of the mean as

$$\left.\frac{dP(z)}{dz}\right|_{z=1}$$

causes the same difficulties as were alluded to earlier. It can be done, but some care is needed with the algebraic manipulation. As before, we proceed to expand $P(z)$ as a power series in y ($= z - 1$) where $z = 1 + y$. Then in terms of y,

$$P(z) = \frac{y(1 - z_1 + y)}{2(1 - z_1)[(1 + y)^2 e^{-3y/2} - 1]}$$

$$= \frac{\dfrac{1}{2} y\left(1 + \dfrac{y}{1 - z_1}\right)}{(1 + 2y + y^2)\left(1 - \dfrac{3}{2} y + \dfrac{9}{8} y^2 - \cdots\right) - 1}$$

$$
= \frac{\dfrac{1}{2}y\left(1 + \dfrac{y}{1 - z_1}\right)}{2y - \dfrac{3}{2}y + y^2\left(1 - 3 + \dfrac{9}{8}\right) + \cdots}
$$

$$
= \frac{1 + \dfrac{y}{1 - z_1}}{1 - \dfrac{7}{4}y + \cdots}
$$

$$
= 1 + \frac{y}{1 - z_1} + \frac{7}{4}y + \cdots
$$

$$
= 1 + y\frac{11 - 7z_1}{4(1 - z_1)} + \cdots
$$

up to terms in y.

Thus the coefficient of y in this expansion is $(11 - 7z_1)/4(1 - z_1)$. This also gives the expected number of passengers waiting when a chair arrives. When $z_1 = -0.36$ its numerical value is $13.52/5.44 = 2.485$.

8
Transient solutions

8.1 The M/M/1 queue

Through equations (2.2) we were able to derive the transient solution for a very simple queueing system. This gives the probabilities for the various states as functions of time. In this particular case we had to solve a pair of differential equations subject to certain initial conditions.

In general the problem of finding the transient solution is not a trivial one. For the single-server queue, with a FIFO queue discipline, random arrivals at rate α and negative exponential service at rate β, where $p_n(t)$ denotes the probability of n customers in the system at time t, the equations to be solved for $p_n(t)$ take the form (2.4):

$$\frac{dp_0(t)}{dt} = -\alpha p_0(t) + \beta p_1(t)$$

$$\frac{dp_1(t)}{dt} = \alpha p_0(t) - (\alpha + \beta)p_1(t) + \beta p_2(t)$$

(8.1)

$$\frac{dp_n(t)}{dt} = \alpha p_{n-1}(t) - (\alpha + \beta)p_n(t) + \beta p_{n+1}(t) \qquad \text{for } n \geq 1$$

etc.

Now this might be regarded as just about the simplest queueing system and yet the equations are very difficult to handle. We quote without proof the solution which corresponds to an empty system at time zero.

$$p_n(t) = e^{-(\alpha + \beta)t}\left[\left(\frac{\alpha}{\beta}\right)^{n/2} I_n(2\sqrt{\alpha\beta}t) + \left(\frac{\alpha}{\beta}\right)^{(n-1)/2} I_{n+1}(2\sqrt{\alpha\beta}t)\right.$$

$$\left. + \left(1 - \frac{\alpha}{\beta}\right)\left(\frac{\alpha}{\beta}\right)^n \sum_{k=n+2}^{\infty} \left(\frac{\alpha}{\beta}\right)^{-k/2} I_k(2\sqrt{\alpha\beta}t)\right]$$

(8.2)

where $I_j(\cdot)$ is a Bessel function. This is a formidable formula, not easy to evaluate, and we shall say no more about it.

8.2 A solution method using generating functions

Consider the birth–death equations (3.1) with $\alpha_n = \alpha$, $\beta_n = n\beta$, i.e. random arrivals at rate α and self-service with a negative exponential distribution at rate β for each customer. Suppose at time $t = 0$ there are i customers in the system. Then the differential equations to be solved are (3.1)

$$\frac{dp_0(t)}{dt} = -\alpha p_0(t) + \beta p_1(t)$$

$$\frac{dp_n(t)}{dt} = \alpha p_{n-1}(t) - (\alpha + n\beta)p_n(t) + (n+1)\beta p_{n+1}(t) \qquad \text{for } n \geq 1$$

(8.3)

If we define the generating function

$$\pi(z, t) = \sum_{n=0}^{\infty} p_n(t)z^n$$

then on multiplication of our equations by z^0, z^1, z^2, etc. followed by addition we obtain

$$\sum_{n=0}^{\infty} z^n \frac{dp_n(t)}{dt} = \alpha \sum_{n=1}^{\infty} z^n p_{n-1}(t) - \alpha \sum_{n=0}^{\infty} z^n p_n(t) - \beta \sum_{n=0}^{\infty} nz^n p_n(t)$$

$$+\beta \sum_{n=0}^{\infty} (n+1)z^n p_{n+1}(t)$$

$$\therefore \quad \frac{\partial \pi(z, t)}{\partial t} = -\alpha(1 - z)\pi(z, t) + \beta(1 - z) \frac{\partial \pi(z, t)}{\partial z}$$

Thus the set of equations for the $p_n(t)$ have been transformed into one equation for $\pi(z, t)$, namely

$$\frac{\partial \pi(z, t)}{\partial t} - \beta(1 - z) \frac{\partial \pi(z, t)}{\partial z} = -\alpha(1 - z)\pi(z, t)$$

(8.4)

The equation above is a special case of Lagrange's equation

$$P \frac{\partial \pi(z, t)}{\partial t} + Q \frac{\partial \pi(z, t)}{\partial z} = R$$

(8.5)

where P, Q and R are functions of π, z and t. If we compare this equation with the general result

$$\frac{\partial \pi}{\partial t} dt + \frac{\partial \pi}{\partial z} dz = d\pi$$

we obtain the subsidiary equations

$$\frac{dt}{P} = \frac{dz}{Q} = \frac{d\pi}{R}$$

(8.6)

which have *two* independent solutions

$$F_1(\pi, z, t) = a$$

$$F_2(\pi, z, t) = b$$

where a and b are constants (of integration).

We do not have time or space to go into the details of the solution of partial differential equations. Suffice it to say that the general solution for this type of equation is given by

$$F_2 = \varphi(F_1) \tag{8.7}$$

where $\varphi(\cdot)$ is some arbitrary function which will be determined by the initial conditions. Thus in our problem the subsidiary equations become

$$\frac{dt}{1} = \frac{dz}{-\beta(1-z)} = \frac{d\pi}{-\alpha(1-z)\pi}$$

Now

$$\frac{dt}{1} = \frac{dz}{\beta(1-z)}$$

gives

$$\frac{dz}{dt} = -\beta(1-z)$$

which in turn gives

$$F_1(\pi, z, t) = (1-z)e^{-\beta t} = a \qquad \text{(a constant)}$$

Similarly

$$\frac{dz}{\beta(1-z)} = \frac{d\pi}{\alpha(1-z)\pi}$$

gives

$$\frac{d\pi}{dz} = \frac{\alpha}{\beta}\pi$$

i.e. $\pi = b \exp\{\alpha z/\beta\}$, i.e.

$$F_2(\pi, z, t) = \pi e^{-\alpha z/\beta} = b \qquad \text{(a constant)}$$

Thus the general solution for π is given by (8.7)

$$\pi(z, t)e^{-\alpha z/\beta} = \varphi[(1-z)e^{-\beta t}]$$

i.e.

$$\pi(z, t) = e^{\alpha z/\beta}\varphi[(1-z)e^{-\beta t}]$$

Now when $t = 0$, $\pi(z, 0) = z^i$ ($p_i(0) = 1$, $p_j(0) = 0$, $j \neq i$)

$$\therefore \quad z^i = e^{\alpha z/\beta}\varphi(1-z)$$

Thus with $1 - z = y$, i.e. $z = 1 - y$

$$\varphi(y) = e^{-\alpha(1-y)/\beta}(1-y)^i$$

and this gives the form of the function $\varphi(\cdot)$ in this case. Thus

$$\pi(z, t) = e^{\alpha z/\beta}e^{-\alpha[1-(1-z)e^{-\beta t}]/\beta}\{1 - (1-z)e^{-\beta t}\}^i$$

$$= \exp\left[-\frac{\alpha}{\beta}(1-z)\left(1 - e^{-\beta t}\right)\right]\left\{1 - (1-z)e^{-\beta t}\right\}^i \tag{8.8}$$

This again is not exactly easy to handle. $p_n(t)$ is given by the coefficient of z^n in the power series expansion of $\pi(z, t)$. Alternatively

$$p_n(t) = \frac{1}{n!} \left. \frac{\partial^n \pi}{\partial z^n} \right|_{z=0}$$

and if we use Leibniz's result for the nth derivative of a product

$$p_n(t) = \frac{1}{n!} \sum_{k=0}^{n} \binom{n}{k} \left[\frac{\alpha}{\beta} \left(1 - e^{-\beta t} \right) \right]^{n-k} \exp\left\{ -\frac{\alpha}{\beta} \left(1 - e^{-\beta t} \right) \right\}$$

$$\times \binom{i}{k} k! e^{-k\beta t} (1 - e^{-\beta t})^{i-k}$$

$$\therefore \; p_n(t) = \frac{1}{n!} \exp\left\{ -\frac{\alpha}{\beta} \left(1 - e^{-\beta t} \right) \right\} \sum_{k=0}^{n} \binom{n}{k} \left(\frac{\alpha}{\beta} \right)^{n-k} (1 - e^{-\beta t})^{n-2k+i} e^{-k\beta t} \binom{i}{k} k! \quad (8.9)$$

This is another somewhat intimidating formula. [Note $\binom{i}{k} = 0$ if $k > i$.] As $t \to \infty$, $e^{-\beta t} \to 0$ and so we can investigate the steady-state solution

$$p_n = \underset{t \to \infty}{\text{Limit}} \; p_n(t)$$

The only term to survive in the sum is the one with $k = 0$. Hence

$$p_n = \underset{t \to \infty}{\text{Limit}} \; p_n(t) = \frac{1}{n!} e^{-\alpha/\beta} \left(\frac{\alpha}{\beta} \right)^n$$

This is in accord with the steady-state solution given in Chapter 3 (a queue with self-service; the M/M/∞ queue). To obtain the steady-state solution by investigating the limit of the transient solution is the hard way to do it! The derivation of the steady-state probabilities can be taken back to an earlier stage. We investigate $\text{Limit}_{t \to \infty} \, \pi(z, t) = \pi(z)$, say; this will form the generating function for the p_n

$$\pi(z) = \sum_{n=0}^{\infty} p_n z^n$$

Thus since $e^{-\beta t} \to 0$ as $t \to \infty$

$$\pi(z) = e^{\alpha z/\beta} e^{-\alpha/\beta} \{1\}^i = e^{\alpha z/\beta} e^{-\alpha/\beta}$$

$$\therefore \; \pi(z) = e^{-\alpha/\beta} \left(1 + \frac{\alpha}{\beta} z + \frac{1}{2!} \left(\frac{\alpha}{\beta} \right)^2 z^2 + \cdots \right)$$

Thus

$$p_n = \text{coefficient of } z^n = \frac{e^{-\alpha/\beta} \left(\dfrac{\alpha}{\beta} \right)^n}{n!}$$

as before. Of course it is independent of i, the initial situation.

We can of course use the generating function $\pi(z, t)$ to obtain an expression for the mean number in the system at time t which we denote by $L(t)$. Since

$$\pi(z, t) = \sum_{n=0}^{\infty} p_n(t)z^n$$

$$\frac{\partial \pi}{\partial z} = \sum_{n=0}^{\infty} np_n(t)z^{n-1}$$

so that

$$\frac{\partial \pi}{\partial z}\bigg|_{z=1} = \sum_{n=0}^{\infty} np_n(t) = L(t) \tag{8.10}$$

This is much easier than using the expression for $p_n(t)$ and attempting the direct summation of the series $\Sigma np_n(t)$.

$$\pi(z, t) = \exp\left[-\frac{\alpha}{\beta}(1-z)(1-e^{-\beta t})\right]\{1-(1-z)e^{-\beta t}\}^i$$

$$\therefore \frac{\partial \pi}{\partial z} = \frac{\alpha}{\beta}(1-e^{-\beta t})\exp\left[-\frac{\alpha}{\beta}(1-z)(1-e^{-\beta t})\right]\left\{1-(1-z)e^{-\beta t}\right\}^i$$

$$+ ie^{-\beta t}\exp\left[-\frac{\alpha}{\beta}(1-z)(1-e^{-\beta t})\right]\{1-(1-z)e^{-\beta t}\}^{i-1}$$

$$\therefore \frac{\partial \pi}{\partial z}\bigg|_{z=1} = \frac{\alpha}{\beta}(1-e^{-\beta t}) + ie^{-\beta t} = L(t) \tag{8.11}$$

Naturally, $\text{Limit}_{t\to\infty} L(t) = \alpha/\beta$ corresponds to the steady-state mean (see Chapter 3) and is independent of i.

Example 8.1

A model for the reproduction by binary fusion of bacteria is as follows: growth of the cell involves replication of N genes within the cell. When all N genes have been replicated, the cell divides and the process repeats itself.

Let $X(t)$ be the number of genes which have been replicated in a cell of age t and let $p_n(t) = \Pr[X(t) = n]$. On the assumption that replication of each gene is a random process which proceeds at an average rate λ, show that

$$\frac{dp_n(t)}{dt} = -\lambda(N-n)p_n(t) + \lambda(N-n+1)p_{n-1}(t), \qquad 1 \le n \le N$$

If $p_n(0) = 1$, $n = 0$, and $p_n(0) = 0$, $n > 0$ show that

$$p_n(t) = \binom{N}{n}e^{-N\lambda t}(e^{\lambda t} - 1)^n$$

Hence find the density function for the random variable τ, the age of a cell when it divides.

In terms of queueing theory this is analogous to a pure arrival process, an arrival being the replication of another gene. When they have all 'arrived', the cell divides and the process begins again.

$$p_n(t + \delta t) = p_n(t)[1 - \lambda(N - n) \, \delta t] + p_{n-1}(t)\lambda(N - n + 1) \, \delta t$$

Since when n genes have replicated, $(N - n)$ genes have yet to replicate and this each one will do in the interval $(t, t + \delta t)$ with probability $\lambda \delta t$ to first order in δt. Thus

$$\frac{p_n(t + \delta t) - p_n(t)}{\delta t} = \lambda(N - n + 1)p_{n-1}(t) - \lambda(N - n)p_n(t)$$

so that as $\delta t \rightarrow 0$ then for $1 \leqslant n \leqslant N$

$$\frac{dp_n(t)}{dt} = \lambda(N - n + 1)p_{n-1}(t) - \lambda(N - n)p_n(t)$$

When $n = 0$ we simply have

$$\frac{dp_0(t)}{dt} = -\lambda N p_0(t)$$

We define the generating function

$$\pi(z, t) = \sum_{n=0}^{N} p_n(t)z^n$$

We multiply the nth equation by z^n and sum to obtain

$$\sum_{n=0}^{N} z^n \frac{dp_n(t)}{dt} = \lambda N \sum_{n=1}^{N} z^n p_{n-1}(t) - \lambda \sum_{n=1}^{N} (n - 1)z^n p_{n-1}(t)$$

$$- \lambda N \sum_{n=0}^{N} z^n p_n(t) + \lambda \sum_{n=0}^{N} n \, z^n p_n(t)$$

$$\therefore \quad \frac{\partial \pi}{\partial t} = \lambda N z[\pi - z^N p_N(t)] - \lambda z^2 \sum_{l=0}^{N-1} l z^{l-1} p_l(t) - \lambda N \pi + \lambda z \frac{\partial \pi}{\partial z}$$

$$\therefore \quad \frac{\partial \pi}{\partial t} = \lambda N(z - 1)\pi - \lambda N z^{N+1} p_N(t) + \lambda z \frac{\partial \pi}{\partial z}$$

$$- \lambda z^2 \left[\frac{\partial \pi}{\partial z} - N z^{N-1} p_N(t) \right]$$

$$\therefore \quad \frac{\partial \pi}{\partial t} = \lambda N(z - 1)\pi - \lambda z(z - 1) \frac{\partial \pi}{\partial z}$$

and we need to solve this partial differential equation subject to the condition $\pi(z, 0) = 1$.

The related equations are

$$\frac{dt}{1} = \frac{dz}{\lambda z(z - 1)} = \frac{d\pi}{\lambda N(z - 1)\pi}$$

The first

$$\frac{dt}{1} = \frac{dz}{\lambda z(z-1)} = \frac{dz}{\lambda} \left[\frac{1}{z-1} - \frac{1}{z} \right]$$

gives

$$\lambda t = \ln A \left(\frac{z-1}{z} \right)$$

i.e.

$$u(z, t, \pi) = \frac{e^{\lambda t} z}{z-1} = c_1 \qquad \text{(a constant)}$$

The second

$$\frac{dz}{\lambda z(z-1)} = \frac{d\pi}{\lambda N(z-1)\pi}$$

gives

$$\frac{dz}{z} = \frac{d\pi}{N\pi}$$

i.e.

$$v(\pi, z, t) = \frac{\pi}{z^N} = c_2 \qquad \text{(a constant)}$$

Thus $v = g(u)$ gives

$$\pi(z, t) = z^N g \left(\frac{z e^{\lambda t}}{z-1} \right)$$

$$\therefore \ \pi(z, 0) = 1 = z^N g \left(\frac{z}{z-1} \right)$$

$$\therefore \ g \left(\frac{z}{z-1} \right) = \frac{1}{z^N}$$

Now with $y = z/(z-1)$, so that $z = y/(y-1)$

$$g(y) = \left(\frac{y-1}{y} \right)^N$$

$$\therefore \ \pi(z, t) = z^N \left[\frac{\dfrac{z e^{\lambda t}}{z-1} - 1}{\dfrac{z e^{\lambda t}}{z-1}} \right]^N$$

$$\therefore \ \pi(z, t) = e^{-N\lambda t} [1 + z(e^{\lambda t} - 1)]^N$$

$p_n(t)$ = coefficient of z^n in this expression,

$$\therefore \ p_n(t) = e^{-N\lambda t}\binom{N}{n}(e^{\lambda t} - 1)^n$$

Thus

$$p_N(t) = e^{-N\lambda t}(e^{\lambda t} - 1)^N = (1 - e^{-\lambda t})^N$$

is the probability that all N genes have replicated by time t and this is necessary for the cell to divide. Thus if T represents the time to replication

$$\Pr(T \leqslant \tau) = p_N(\tau) = (1 - e^{-\lambda \tau})^N$$

is the distribution function for the time to divide. By differentiation, the probability density function is

$$\lambda N e^{-\lambda \tau}(1 - e^{-\lambda \tau})^{N-1}$$

8.3 Examples 15

1 Use the equations (8.3), namely

$$\frac{dp_0(t)}{dt} = -\alpha p_0(t) + \beta p_1(t)$$

$$\frac{dp_n(t)}{dt} = \alpha p_{n-1}(t) - (\alpha + n\beta)p_n(t) + (n+1)\beta p_{n+1}(t) \qquad \text{for } n \geqslant 1$$

to show that if

$$L(t) = \sum_{n=0}^{\infty} np_n(t)$$

then

$$\frac{dL(t)}{dt} = \alpha - \beta L(t)$$

Deduce that if initially there are i customers in the system then

$$L(t) = \frac{\alpha}{\beta}(1 - e^{-\beta t}) + ie^{-\beta t}$$

2 Rewrite the generating function (8.8) in the form

$$\pi(z, t) = e^{-\alpha(1 - e^{-\beta t})/\beta}e^{\alpha z(1 - e^{-\beta t})/\beta}(1 - e^{-\beta t} + ze^{-\beta t})^i$$

Show that

$$(1 - e^{-\beta t} + ze^{-\beta t})^i = \sum_{k=0}^{i}\binom{i}{k}e^{-k\beta t}(1 - e^{-\beta t})^{i-k}z^k$$

and that

$$e^{az(1-e^{-\beta t})/\beta} = \sum_{m=0}^{\infty} \left[\frac{a}{\beta} (1 - e^{-\beta t})^m \right] \frac{z^m}{m!}$$

Deduce the form of $p_n(t)$ as given by (8.9).

3 A telephone exchange has N lines. On each line the interval between incoming calls has a negative exponential distribution with mean $1/a$ and the duration of calls has a negative exponential distribution with mean $1/\beta$. Calls which arrive when all lines are busy are lost to the system. If $p_n(t)$ denotes the probability that n lines are busy at time t, show that

$$\frac{dp_n(t)}{dt} = (N - n + 1)ap_{n-1}(t) - [(N - n)a + n\beta]p_n(t) + (n + 1)\beta p_{n+1}(t)$$

$$1 \le n \le N - 1$$

and obtain equations for $dp_0(t)/dt$ and $dp_N(t)/dt$.
If $\pi(z, t) = \sum_{n=0}^{N} p_n(t)z^n$ show that

$$\frac{\partial \pi(z, t)}{\partial t} - (1 - z)(\beta + az) \frac{\partial \pi(z, t)}{\partial z} + Na(1 - z)\pi(z, t) = 0$$

Show that the solution corresponding to no lines being busy initially is

$$\pi(z, t) = \left\{ \frac{(\beta + ae^{-(a+\beta)t}) + az(1 - e^{-(a+\beta)t})}{a + \beta} \right\}^N$$

and hence find an expression for the mean number of busy lines at time t.

4 For the machine interference problem (Section 3.7) there are N identical machines. Each machine breaks down at random in running time at an average rate a. The time taken by an operative to repair a broken machine has a negative exponential distribution with mean $1/\beta$. Assume that there are N operatives who work as a team looking after the N machines. If $p_n(t)$ denotes the probability that n machines are stopped at time t, show that the equations which determine the $p_n(t)$ are the same as those obtained in question 3 of these examples.

The $\pi(z, t)$ corresponds to the situation where all the machines are running initially. Deduce that in this case

$$p_n(t) = \binom{N}{n} \left[\frac{a}{a + \beta} (1 - e^{-(a+\beta)t}) \right]^n \left\{ \frac{\beta + ae^{-(a+\beta)t}}{a + \beta} \right\}^{N-n}$$

Can you use the results given by (2.2), with $p_R(0) = 1$ and $p_S(0) = 0$, to obtain this expression by an elementary argument? Show that

$$\text{Limit}_{t \to \infty} \pi(z, t) = \left(\frac{\beta + az}{a + \beta} \right)^N$$

and hence find the steady-state distribution of the number of stopped machines.

8.4   Solutions to Examples 15

1 $p_0'(t) = -\alpha p_0(t) + \beta p_1(t)$
$p_1'(t) = \alpha p_0(t) - (\alpha + \beta)p_1(t) + 2\beta p_2(t)$
$p_2'(t) = \alpha p_1(t) - (\alpha + 2\beta)p_2(t) + 3\beta p_3(t)$ etc.

Multiply these equations in turn by 0, 1, 2, 3,..., etc. and add

$$\therefore \sum_{n=0}^{\infty} n p_n'(t) = \alpha \sum_{n=0}^{\infty} (n+1)p_n(t) - \alpha \sum_{n=0}^{\infty} n p_n(t) - \beta \sum_{n=1}^{\infty} n^2 p_n(t)$$

$$+ \beta \sum_{n=1}^{\infty} n(n-1)p_n(t)$$

$$\therefore \frac{dL(t)}{dt} = \alpha \sum_{n=0}^{\infty} p_n(t) - \beta \sum_{n=1}^{\infty} n p_n(t)$$

$$= \alpha - \beta L(t) \qquad\qquad \text{i.e. } dL/dt + \beta L = \alpha$$

This is a first-order linear differential equation for $L(t)$. A particular integral is given by $L = \alpha/\beta$. The complementary function takes the form $Ae^{-\beta t}$

$$\therefore L(t) = \frac{\alpha}{\beta} + Ae^{-\beta t}$$

When $t = 0$, $L(0) = i$ and so

$$i = \frac{\alpha}{\beta} + A$$

$$\therefore A = i - \frac{\alpha}{\beta}$$

$$\therefore L(t) = \frac{\alpha}{\beta}(1 - e^{-\beta t}) + ie^{-\beta t}$$

2 $$\pi(z, t) = e^{-\alpha(1 - e^{-\beta t})/\beta} e^{\alpha z(1 - e^{-\beta t})/\beta}(1 - e^{-\beta t} + ze^{-\beta t})^i$$

By the binomial expansion

$$(1 - e^{-\beta t} + ze^{-\beta t})^i = \sum_{k=0}^{i} \binom{i}{k} e^{-k\beta t}(1 - e^{-\beta t})^{i-k} z^k$$

$$e^{\alpha z(1 - e^{-\beta t})/\beta} = \sum_{m=0}^{\infty} \frac{\left(\dfrac{\alpha}{\beta}\right)^m (1 - e^{-\beta t})^m z^m}{m!} = \sum_{n=k}^{\infty} \frac{\left(\dfrac{\alpha}{\beta}\right)^{n-k} (1 - e^{-\beta t})^{n-k} z^{n-k}}{(n-k)!}$$

Thus in the product, and including $\exp[-\alpha(1 - e^{-\beta t})/\beta]$, the coefficient of z^n is given by

$$p_n(t) = \exp\left[-\frac{\alpha}{\beta}(1 - e^{-\beta t})\right] \sum_{k=0}^{n} \binom{i}{k} e^{-k\beta t}(1 - e^{-\beta t})^{i-k}\left(\frac{\alpha}{\beta}\right)^{n-k} \frac{(1 - e^{-\beta t})^{n-k}}{(n-k)!}$$

$$= \exp\left[-\frac{\alpha}{\beta}(1 - e^{-\beta t})\right] \sum_{k=0}^{n} \binom{i}{k} e^{-k\beta t}(1 - e^{-\beta t})^{n-2k+i} \frac{n!}{k!(n-k)!} \frac{k!}{n!}$$

$$= \frac{1}{n!} \exp\left[-\frac{\alpha}{\beta}(1 - e^{-\beta t})\right] \sum_{k=0}^{n} \binom{n}{k}\left(\frac{\alpha}{\beta}\right)^{n-k}(1 - e^{-\beta t})^{n-2k+i} e^{-k\beta t} k! \binom{i}{k}$$

as required; where again $\binom{i}{k}$ for $k > i$ is taken to be zero.

3 $p_0(t + \delta t) = p_0(t)(1 - N\alpha\delta t) + p_1(t)\beta\delta t$

$p_n(t + \delta t) = (N - n + 1)\alpha\delta t\, p_{n-1}(t)$
$\qquad + p_n(t)(1 - (N - n)\alpha\delta t)(1 - n\beta\delta t) + p_{n+1}(t)(n + 1)\beta\delta t$

for $1 \leq n \leq N - 1$

$p_N(t + \delta t) = \alpha\delta t\, p_{N-1}(t) + p_N(t)(1 - N\beta\delta t)$

When n lines are busy new calls can arrive only on the $(N - n)$ free lines and at an average rate α on each of them; calls can finish at rate $n\beta$ (at rate β on each of the n busy lines). We have worked to first-order terms in δt. Thus

$p_0'(t) = -N\alpha p_0(t) + \beta p_1(t)$
$p_1'(t) = [N - 0]\alpha p_0(t) - [(N - 1)\alpha + \beta]p_1(t) + 2\beta p_2(t)$
$p_2'(t) = [N - 1]\alpha p_1(t) - [(N - 2)\alpha + 2\beta]p_2(t) + 3\beta p_3(t)$

$p_{N-1}'(t) = [N - (N - 2)]\alpha p_{N-2}(t) - [(N - (N - 1))\alpha + (N - 1)\beta]p_{N-1}(t) + N\beta p_N(t)$
$p_N'(t) = [N - (N - 1)]\alpha p_{N-1}(t) - [(N - N)\alpha + N\beta]p_N(t)$
$\qquad 0 = [N - N]\alpha p_N(t)$

Note the last equation which says $0 = 0$. Multiply these equations in turn by $z^0, z^1, z^2, ..., z^N$ and z^{N+1} and add to give

$$\frac{\partial \pi}{\partial t} = N\alpha z \sum_{n=0}^{N} z^n p_n(t) - \alpha z^2 \sum_{n=0}^{N} nz^{n-1}p_n(t) - N\alpha \sum_{n=0}^{N} z^n p_n(t) + \alpha z \sum_{n=0}^{N} nz^{n-1}p_n(t)$$

$$- \beta z \sum_{n=0}^{N} nz^{n-1}p_n(t) + \beta \sum_{n=0}^{N} nz^{n-1}p_n(t)$$

Note that $np_n(t)z^{n-1}$ is zero when $n = 0$. It has been inserted in the last three sums. Thus

$$\frac{\partial \pi}{\partial t} = N\alpha z\pi - \alpha z^2 \frac{\partial \pi}{\partial z} - N\alpha\pi + \alpha z \frac{\partial \pi}{\partial z} - \beta z \frac{\partial \pi}{\partial z} + \beta \frac{\partial \pi}{\partial z}$$

$$\therefore \quad \frac{\partial \pi}{\partial t} - (1 - z)(\alpha z + \beta) \frac{\partial \pi}{\partial z} + N\alpha(1 - z)\pi = 0$$

The subsidiary equations are:

$$\frac{dt}{1} = \frac{dz}{(z-1)(\beta + az)} = \frac{d\pi}{N\alpha(z-1)\pi}$$

which gives

$$\int dt = -\frac{1}{\alpha + \beta} \int \left(\frac{\alpha}{az + \beta} - \frac{1}{z-1} \right) dz$$

$$\therefore \quad -(\alpha + \beta)t = \ln\left[\left(\frac{az + \beta}{z-1} \right) a \right]$$

whence

$$F_1(\pi, z, t) = \left(\frac{z-1}{\beta + az} \right) e^{-(\alpha + \beta)t} = a$$

Also

$$N \int \frac{\alpha dz}{az + \beta} = \int \frac{d\pi}{\pi}$$

whence

$$\ln b(az + \beta)^N = \ln \pi$$

i.e.

$$F_2(\pi, z, t) = \frac{\pi}{(az + \beta)^N} = b$$

Thus the solution of our partial differential equation takes the form

$$\pi(z, t) = (az + \beta)^N \varphi\left[\frac{(z-1)e^{-(\alpha + \beta)t}}{az + \beta} \right]$$

When $t = 0$, $\pi(z, 0) = 1$

$$\therefore \quad \varphi\left(\frac{z-1}{az + \beta} \right) = (az + \beta)^{-N}$$

If

$$\frac{z-1}{az + \beta} = s$$

then

$$z = \frac{1 + \beta s}{1 - \alpha s}$$

and

$$\beta + az = \frac{\alpha + \beta}{1 - as}$$

so that

$$\varphi(s) = \left(\frac{1 - as}{\alpha + \beta}\right)^N$$

$$\therefore \ \pi(z, t) = (\beta + az)^N \left[\frac{1 - \dfrac{\alpha(z - 1)e^{-(\alpha + \beta)t}}{az + \beta}}{\alpha + \beta} \right]^N$$

$$\therefore \ \pi(z, t) = \left[\frac{az + \beta - \alpha(z - 1)e^{-(\alpha + \beta)t}}{\alpha + \beta} \right]^N$$

$$\therefore \ \pi(z, t) = \left[\frac{\beta + ae^{-(\alpha + \beta)t} + z(\alpha - ae^{-(\alpha + \beta)t})}{\alpha + \beta} \right]^N$$

The expected number of busy lines at time t is given by $(\partial\pi/\partial z)|_{z=1}$

$$\frac{\partial\pi(z, t)}{\partial z} = N\left[\frac{\beta + ae^{-(\alpha + \beta)t} + z(\alpha - ae^{-(\alpha + \beta)t})}{\alpha + \beta} \right]^{N-1} \cdot \frac{(\alpha - ae^{-(\alpha + \beta)t})}{\alpha + \beta}$$

$$\therefore \frac{\partial\pi}{\partial z}\bigg|_{z=1} = \frac{N(\alpha - a^{-(\alpha + \beta)t})}{\alpha + \beta}$$

4 Essentially we have each machine being looked after by its own operator. Thus when n machines are stopped the breakdown rate (rate of new arrivals) is $(N - n)a$ and the repair rate is $n\beta$ (n of the operators will be working on repairs).

Thus the equations for the $p_n(t)$ are identical to those in question 3. The initial situation is also the same since $p_0(0) = 1$ and $p_n(0) = 0$. Thus as before

$$\pi(z, t) = \left[\frac{\beta + ae^{-(\alpha + \beta)t}}{\alpha + \beta} + \frac{z(\alpha - ae^{-(\alpha + \beta)t})}{\alpha + \beta} \right]^N$$

and the coefficient of z^n in this binomial expansion is

$$p_n(t) = \binom{N}{n} \frac{a^n(1 - e^{-(\alpha + \beta)t})^n(\beta + ae^{-(\alpha + \beta)t})^{N-n}}{(\alpha + \beta)^N}$$

From (2.2), where one machine is looked after by one operator, the probability that that machine (it could be any of them) is stopped at time t is

$$(p_s(t)) \ \frac{\alpha - ae^{-(\alpha + \beta)t}}{\alpha + \beta}$$

and the probability that it is running at time t is

$$1 - \frac{\alpha - \alpha e^{-(\alpha + \beta)t}}{\alpha + \beta} = \frac{\beta + \alpha e^{-(\alpha + \beta)t}}{\alpha + \beta}$$

Thus the number of machines which are stopped at time t has a binomial distribution with parameters N and

$$\frac{\alpha(1 - e^{-(\alpha + \beta)t})}{\alpha + \beta}$$

Hence the expression for $p_n(t)$ follows. Since $\text{Limit}_{t \to \infty}\, e^{-(\alpha + \beta)t} = 0$,

$$\text{Limit}_{t \to \infty} \pi(z, t) = \left(\frac{\beta + \alpha z}{\alpha + \beta}\right)^N$$

$$= \sum_{n=0}^{N} \binom{N}{n} \left(\frac{\beta}{\alpha + \beta}\right)^{N-n} \left(\frac{\alpha}{\alpha + \beta}\right)^n z^n$$

hence

$$p_n = \text{coeff. of } z^n = \binom{N}{n} \left(\frac{\beta}{\alpha + \beta}\right)^{N-n} \left(\frac{\alpha}{\alpha + \beta}\right)^n$$

which is the corresponding distribution in the steady-state situation ($p_S = \alpha/(\alpha + \beta)$, $p_R = \beta/(\alpha + \beta)$).

9
Networks of queues

9.1 Introduction

To date we have looked at a queueing system more or less in isolation. However, in some situations the output from one queue can form the input to another, or the outputs from two or more systems may combine to form the input to another, and the output from this latter queueing system can split to form inputs to other systems, etc. Such queueing networks arise quite naturally in industrial production processes where part-finished items have to undergo a number of processes before they are ready for distribution to consumers. Another context in which queueing networks arise is that of electronic communication and transmission networks. Messages for transmission are coded into electronic signals and these signals have to be 'processed' at the transmitter, at points in the transmission system, whatever that is, and again at the receiving and decoding apparatus.

A fundamental result which we need to be quite clear about is that the sum of two independent Poisson streams is also a Poisson stream. If $X(t)$ represents the number of arrivals by time t from a Poisson stream with average arrival rate α, and $Y(t)$ represents the number of arrivals by time t from an independent Poisson stream with average arrival rate β, and $Z(t) = X(t) + Y(t)$, then $Z(t)$ is the number of arrivals by time t from a Poisson stream with average arrival rate $\alpha + \beta$.

$$\Pr(X(t) = x) = \frac{(\alpha t)^x e^{-\alpha t}}{x!} \tag{9.1}$$

$$\Pr(Y(t) = y) = \frac{(\beta t)^y e^{-\beta t}}{y!} \tag{9.2}$$

Thus

$$\Pr(Z(t) = z) = \sum_{x=0}^{z} \Pr(X(t) = x)\Pr(Y(t) = z - x)$$

$$= \sum_{x=0}^{z} \frac{(\alpha t)^x e^{-\alpha t}}{x!} \frac{(\beta t)^{z-x} e^{-\beta t}}{(z-x)!}$$

$$\therefore \ \Pr(Z(t) = z) = \frac{e^{-(\alpha+\beta)t}}{z!} \sum_{x=0}^{z} \binom{z}{x}(\alpha t)^x (\beta t)^{z-x}$$

$$= \frac{[(\alpha + \beta)t]^z e^{-(\alpha + \beta)t}}{z!} \qquad \text{for } z = 0, 1, 2, \dots \tag{9.3}$$

This result easily extends to the sum of three or more *independent* input streams.

9.2   The output from a queue

It can be shown that departing customers from an M/M/1, an M/M/c and an M/M/∞ queue also form a Poisson stream. We consider here the case of the M/M/c system. The easier case of the M/M/1 system, where an alternative derivation of the result is possible, is left as an exercise.

For the M/M/c system where arrivals occur at random at rate α, and service has a negative exponential distribution at rate β from each of c servers, the steady-state probability that there are n customers in the system, p_n, is derived from the solution of the birth–death equations (3.3) and (3.4) with $\alpha_n = \alpha$ and $\beta_n = n\beta$ ($1 \leq n \leq c$), $\beta_n = c\beta$ ($n > c$). Thus the p_n satisfy the recurrence relations

$$p_n = \frac{\alpha}{n\beta} p_{n-1} \qquad \text{for } 1 \leq n \leq c$$

$$p_n = \frac{\alpha}{c\beta} p_{n-1} \qquad \text{for } n > c$$

(9.4)

Of course the p_n describe, in a probability sense, the state of the system at an arbitrary moment in time. In particular, p_n is the probability that a departing customer leaves n customers in the system in the steady state.

Let τ represent an inter-departure interval and let $N(t)$ be the state of the system at time t after a departure. We define

$$S_n(t) = \Pr(N(t) = n \text{ and } \tau > t)$$

so that in the above there have been no further departures by time t. Thus, by considering the possible activity in the interval $(t, t + \delta t)$ in which there can be no departures, we obtain to first order in δt:

$$S_0(t + \delta t) = (1 - \alpha\delta t)S_0(t)$$
$$S_n(t + \delta t) = \alpha\delta t S_{n-1}(t) + (1 - \alpha\delta t)(1 - n\beta\delta t)S_n(t); \qquad n \leq c$$
$$S_n(t + \delta t) = \alpha\delta t S_{n-1}(t) + (1 - \alpha\delta t)(1 - c\beta\delta t)S_n(t); \qquad n > c$$

Thus

$$\frac{dS_0(t)}{dt} = -\alpha S_0(t)$$

$$\frac{dS_n(t)}{dt} = \alpha S_{n-1}(t) - (\alpha + n\beta)S_n(t); \qquad n \leq c$$

(9.5)

$$\frac{dS_n(t)}{dt} = \alpha S_{n-1}(t) - (\alpha + c\beta)S_n(t); \qquad n > c$$

and we have to solve these differential-difference equations for $S_n(t)$ subject to the initial conditions $S_n(0) = p_n$ since there are n in the system at the moment of departure.

It is easy to see that $S_0(t) = p_0 e^{-\alpha t}$ and $S_n(t) = p_n e^{-\alpha t}$ gives the solution that fits the boundary conditions. For direct substitution in the first set of equations gives on the left, $-\alpha p_n e^{-\alpha t}$ and on the right,

$$\alpha p_{n-1} e^{-\alpha t} - \alpha p_n e^{-\alpha t} - n\beta p_n e^{-\alpha t}$$

and this is just $-p_n\alpha e^{-\alpha t}$ since $p_n = \alpha p_{n-1}/(n\beta)$ for $n \leqslant c$. This solution also extends to the second set of equations where $n > c$. Thus

$$S_n(t) = p_n e^{-\alpha t}; \qquad \tau > t \tag{9.6}$$

Thus if we consider all possible states n

$$\Pr(\tau > t) = \sum_{n=0}^{\infty} S_n(t) = e^{-\alpha t} \sum_{n=0}^{\infty} p_n = e^{-\alpha t} \tag{9.7}$$

Thus τ has a negative exponential distribution with probability density function $\alpha e^{-\alpha t}$; $t \geqslant 0$ as required. This shows that the inter-departure times have a negative exponential distribution so that the departures also form a Poisson stream. We can however deduce more, namely that the inter-departure time is independent of the states of the system prior to the moment of departure. For if we consider

$$\Pr(N(t + \delta t) = n \text{ and } t < \tau < t + \delta t) = S_{n+1}(t)(n+1)\beta\delta t \qquad \text{if } n+1 \leqslant c$$
$$= S_{n+1}(t)c\beta\delta t \qquad \text{if } n+1 > c$$

then in both cases because of the relationship between p_{n+1} and p_n

$$\left.\begin{aligned}\Pr(N(\tau +) = n \text{ and } t < \tau < t + \delta t) &= p_{n+1}(n+1)\beta e^{-\alpha t}\,\delta t \\ &= p_{n+1}c\beta e^{-\alpha t}\,\delta t\end{aligned}\right\} = p_n\alpha e^{-\alpha t}\,\delta t \tag{9.8}$$

$$= \Pr(N(\tau +) = n) \times \Pr(t \leqslant \tau \leqslant t + \delta t)$$

and the product form shows that τ and $N(\tau)$ are independent. Thus the number of customers in the system at any time t is not dependent on the state of the system at earlier departure times nor on the earlier inter-departure times.

9.3 Examples 16

1 $X(t)$ has a Poisson distribution with mean αt and $Y(t)$ independent of $X(t)$ has a Poisson distribution with mean βt.

Show that $X(t)$ has probability generating function

$$P_X(s) = \sum_{n=0}^{\infty} Pr(X(t) = n)s^n = e^{-\alpha t(1-s)}$$

and hence show that $Z(t) = X(t) + Y(t)$ has a Poisson distribution with mean $(\alpha + \beta)t$. Refer back to question 3 of Examples 2.

2 (a) If X has a negative exponential distribution with mean $1/\alpha$, and Y has a negative exponential distribution with mean $1/\beta$, where X and Y are independent, then if $Z = X + Y$ show that Z has probability density function

$$f_Z(z) = \frac{\alpha\beta}{\beta - \alpha}e^{-\alpha z} - \frac{\alpha\beta}{\beta - \alpha}e^{-\beta z}$$

(b) Consider an M/M/1 queue with random arrivals at average rate α, and negative exponential service at rate β from one server.

(i) Show that if a customer leaves the system empty, the inter-departure time is the sum of an inter-arrival time and a service time.

(ii) Show that if a customer leaves a busy server the inter-departure time is a service time.

Combine the results from (a) and (b), (i) and (ii), to show that the inter-departure time has a negative exponential distribution with mean $1/\alpha$.

9.4   Solutions to examples 16

1

$$P_X(s) = \sum_{n=0}^{\infty} \frac{e^{-\alpha t}(\alpha t)^n}{n!} s^n = e^{-\alpha t} \sum_{n=0}^{\infty} \frac{(s\alpha t)^n}{n!}$$

$$= e^{-\alpha t(1-s)}$$

Similarly $P_Y(s) = e^{-\beta t(1-s)}$ and since $X(t)$ and $Y(t)$ are independent and $Z(t) = X(t) + Y(t)$

$$P_Z(s) = P_X(s)P_Y(s)$$

$$= e^{-\alpha t(1-s)} \cdot e^{-\beta t(1-s)} = e^{-(\alpha+\beta)t(1-s)}$$

whence the result follows.

2 (a) By the usual convolution argument ((1.7) for example)

$$f_Z(z) = \int_0^z f_X(x)f_Y(z-x)\, dx$$

$$= \int_0^z \alpha e^{-\alpha x}\beta e^{-\beta(z-x)}\, dx$$

$$= \alpha\beta e^{-\beta z} \int_0^z e^{(\beta-\alpha)x}\, dx$$

$$= \frac{\alpha\beta e^{-\beta z}}{\beta-\alpha}\left[e^{(\beta-\alpha)z} - 1\right]$$

$$= \frac{\alpha\beta e^{-\alpha z}}{\beta-\alpha} - \frac{\alpha\beta e^{-\beta z}}{\beta-\alpha}$$

(b) (i) In this case we have to wait until a new arrival has occurred and the service time of that arrival is complete; then the next departure occurs.

(ii) In this case the next customer in the system enters service immediately and at the end of that service time the next departure occurs.

Thus if the inter-departure time has probability density function $\varphi(t)$,

$$\varphi(t) = p_0\left[\frac{\alpha\beta e^{-\alpha t}}{\beta-\alpha} - \frac{\alpha\beta e^{-\beta t}}{\beta-\alpha}\right] + (1 - p_0)\beta e^{-\beta t}$$

where we have used the result from (a). But for the M/M/1 system

$$p_0 = 1 - \frac{\alpha}{\beta} = \frac{\beta-\alpha}{\beta} \quad \text{and} \quad 1 - p_0 = \frac{\alpha}{\beta}$$

$$\therefore \quad \varphi(t) = \alpha e^{-\alpha t} - \alpha e^{-\beta t} + \alpha e^{-\beta t} = \alpha e^{-\alpha t} \qquad \text{for } t \geq 0$$

and this is the required result.

9.5 Two systems in tandem

Example 9.1

Consider the network in Fig. 9.1. Customers arrive at random at rate α to a first system which has negative exponential service at rate β_1 from its one server. After service customers then enter a second system which has one server and negative exponential service at rate β_2. If $p(n_1, n_2)$ denotes the steady-state probability that there are n_1 customers in the first system and n_2 customers in the second system, find $p(n_1, n_2)$ and the average number of customers in each system. We have to assume that $\alpha < \beta_1$ and $\alpha < \beta_2$ for a steady state to exist.

The first system behaves like an M/M/1 system with arrival rate α and service rate β_1. The departures from this system are the arrivals to the second system. But we know that these departures form a Poisson stream with rate α. Thus the second system also behaves like an M/M/1 system with arrival rate α and service rate β_2. However, the two systems are also independent. The departures from the first system form the arrivals to the second system but the state of the first system $(N_1 = n_1)$ is independent of the previous sequence of departures from it and it is these departures which define the arrivals to the second system. Thus if N_1 and N_2 represent the numbers of customers in the two systems in the steady state

$$p(n_1, n_2) = \Pr(N_1 = n_1, N_2 = n_2) = \Pr(N_1 = n_1)\Pr(N_2 = n_2)$$

Thus from our results for the M/M/1 queue (2.6)

$$p(n_1, n_2) = (1 - \rho_1)\rho_1^{n_1}(1 - \rho_2)\rho_2^{n_2}; \qquad n_1, n_2 \geq 0 \tag{9.9}$$

where $\rho_1 = \alpha/\beta_1$ and $\rho_2 = \alpha/\beta_2$ and $0 < \frac{\beta_1}{\beta_2} < 1$.

By summing over n_1 we obtain

$$\Pr(N_2 = n_2) = (1 - \rho_2)\rho_2^{n_2} \quad \text{and} \quad E[N_2] = \frac{\rho_2}{1 - \rho_2} \tag{9.10}$$

Similarly

$$\Pr(N_1 = n_1) = (1 - \rho_1)\rho_1^{n_1} \quad \text{and} \quad E[N_1] = \frac{\rho_1}{1 - \rho_1} \tag{9.11}$$

$$E[N_1 + N_2] = \sum_{n_1}\sum_{n_2}(n_1 + n_2)(1 - \rho_1)\rho_1^{n_1}(1 - \rho_2)\rho_2^{n_2}$$

$$= \sum_{n_1}n_1(1 - \rho_1)\rho_1^{n_1} \cdot \sum_{n_2}(1 - \rho_2)\rho_2^{n_2} + \sum_{n_2}n_2(1 - \rho_2)\rho_2^{n_2}\sum_{n_1}(1 - \rho_1)\rho_1^{n_1}$$

$$= \frac{\rho_1}{1 - \rho_1} + \frac{\rho_2}{1 - \rho_2} = E[N_1] + E[N_2] \tag{9.12}$$

The result $p(n_1, n_2) = \Pr(N_1 = n_1)\Pr(N_2 = n_2)$, the so-called product-form solution, is valid if the two systems are M/M/1, M/M/c or M/M/∞ in any combination. Indeed if

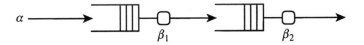

Fig. 9.1

there are *more* than two systems in tandem then the result above easily generalises. See also Examples 17, question 1.

9.6 Splitting a Poisson stream

Suppose that a proportion p of the departures from an M/M/1 system (or an M/M/c or an M/M/∞ system) join a second system, whereas the remainder depart altogether (Fig. 9.2). Then the input to the second system is a Poisson stream with rate $p\alpha$.

Mathematically we are dealing with a situation where, with probability p, a Poisson stream (departure) at rate α becomes an arrival to the second system. Let $X(t)$ denote the number of departures from the first system and $Y(t)$ the number of arrivals to the second system by time t. Then

$$\Pr(Y(t) = m \,|\, X(t) = n) = \binom{n}{m} p^m (1 - p)^{n - m}$$

Thus

$$\Pr(Y(t) = m) = \sum_{n=m}^{\infty} \Pr(Y(t) = m \,|\, X(t) = n) \Pr(X(t) = n)$$

where n has to be at least as large as m.

$$\therefore\ \Pr(Y(t) = m) = \sum_{n=m}^{\infty} e^{-\alpha t} \frac{(\alpha t)^n}{n!} \frac{n!}{m!(n-m)!} p^m (1 - p)^{n-m}$$

$$= \frac{p^m e^{-\alpha t} (\alpha t)^m}{m!} \sum_{n=m}^{\infty} \frac{[\alpha t(1-p)]^{n-m}}{(n-m)!}$$

$$= \frac{(p\alpha t)^m e^{-\alpha t}}{m!} e^{\alpha t(1-p)}$$

$$= \frac{(p\alpha t)^m e^{-p\alpha t}}{m!} \tag{9.13}$$

Thus $Y(t)$ forms a Poisson stream with input rate $p\alpha$.

This result also extends to the case where the original stream splits into three or more independent streams.

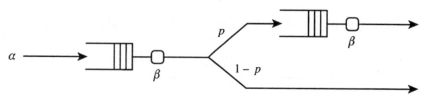

Fig. 9.2

9.7 Feed forward networks

The product-form solution stated in the earlier example will be valid for any 'feed forward' network such as that indicated in the next example. In this network, which has two external inputs (one to each of the waiting systems) it is not possible for any customer to visit any of the systems more than once. Of course all the systems must be such that a steady state exists (Fig. 9.3).

System 1 operates as an M/M/1 system with $\rho_1 = \alpha_1/\beta_1$. System 2 operates as an independent M/M/1 system with $\rho_2 = \frac{1}{3}\,\alpha_1/\beta_2$. System 3 operates as an independent M/M/1 system with $\rho_3 = (\alpha_2 + \frac{2}{3}\alpha_1)/\beta_3$. Thus provided $\max(\rho_1, \rho_2, \rho_3) < 1$

$$\Pr(N_1 = n_1, N_2 = n_2, N_3 = n_3) = (1 - \rho_1)(1 - \rho_2)(1 - \rho_3)\rho_1^{n_1}\rho_2^{n_2}\rho_3^{n_3}; \qquad n_1, n_2, n_3 \geq 0$$

Thus, for example, given that a customer took the route from 1 to 2, his expected time in the system is given by

$$E[W_1] + \frac{1}{\beta_1} + E[W_2] + \frac{1}{\beta_2}$$

where $E[W_i]$ is the expected queueing time at system i. Thus on using the result (2.13) we obtain

$$\frac{\alpha_1}{\beta_1(\beta_1 - \alpha_1)} + \frac{1}{\beta_1} + \frac{\frac{1}{3}\alpha_1}{\beta_2\left(\beta_2 - \frac{1}{3}\alpha_1\right)} + \frac{1}{\beta_2}$$

or

$$\frac{1}{\beta_1 - \alpha_1} + \frac{1}{\beta_2 - \frac{1}{3}\alpha_1}$$

as an alternative form from (2.14).

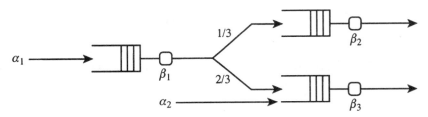

Fig. 9.3

Example 9.2

The network of M/M/1 systems, illustrated in Fig. 9.4, represents a production process for completing the manufacture of certain components. The components arrive at the first station at an average rate α and are processed at rate β_1. Two-thirds of them are

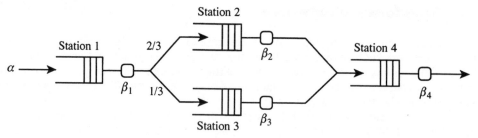

Fig. 9.4

then sent to station 2 where they are modified in a particular way at rate β_2. The remaining third are sent to station 3 where they are modified somewhat differently at rate β_3. Finally, all components receive their final check at station 4 where they are processed at rate β_4. Estimate the mean throughput time for each route.

We let $p(n_1, n_2, n_3, n_4)$ be the probability that there are n_1 components in the system for station 1, n_2 for station 2, etc. Then, using results for the steady-state solution of an M/M/1 system (2.6) along with the product rule, we have

$$p(n_1, n_2, n_3, n_4) = (1 - \rho_1)(1 - \rho_2)(1 - \rho_3)(1 - \rho_4)\rho_1^{n_1}\rho_2^{n_2}\rho_3^{n_3}\rho_4^{n_4}$$

where $\rho_1 = \alpha/\beta_1$, $\rho_2 = (\frac{2}{3}\alpha)/\beta_2$, $\rho_3 = (\frac{1}{3}\alpha)/\beta_3$ and $\rho_4 = \alpha/\beta_4$ and we must have max $(\rho_1, \rho_2, \rho_3, \rho_4) < 1$.

For the components which pass through station 2, the expected time in the system is (we have used (2.13) or (2.14))

$$E[W_1] + \frac{1}{\beta_1} + E[W_2] + \frac{1}{\beta_2} + E[W_4] + \frac{1}{\beta_4}$$

$$= \frac{\alpha}{\beta_1(\beta_1 - \alpha)} + \frac{1}{\beta_1} + \frac{\frac{2}{3}\alpha}{\beta_2\left(\beta_2 - \frac{2}{3}\alpha\right)} + \frac{1}{\beta_2} + \frac{\alpha}{\beta_4(\beta_4 - \alpha)} + \frac{1}{\beta_4}$$

$$= \frac{1}{(\beta_1 - \alpha)} + \frac{1}{\left(\beta_2 - \frac{2}{3}\alpha\right)} + \frac{1}{(\beta_4 - \alpha)}$$

For the components which pass through station 3, the expected time in the system is

$$\frac{1}{(\beta_1 - \alpha)} + \frac{1}{\left(\beta_3 - \frac{1}{3}\alpha\right)} + \frac{1}{(\beta_4 - \alpha)}$$

9.8　Examples 17

1　For the two M/M/1 systems in tandem (Fig. 9.5) show that the equations for the steady-state probabilities

$$\Pr(N_1 = n_1, N_2 = n_2) = p(n_1, n_2)$$

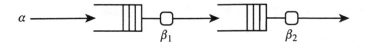

Fig. 9.5

are:

$$\alpha p(0, 0) = \beta_2 p(0, 1)$$
$$(\alpha + \beta_1)p(n_1, 0) = \alpha p(n_1 - 1, 0) + \beta_2 p(n_1, 1); \qquad n_1 > 0$$
$$(\alpha + \beta_2)p(0, n_2) = \beta_1 p(1, n_2 - 1) + \beta_2 p(0, n_2 + 1); \qquad n_2 > 0$$
$$(\alpha + \beta_1 + \beta_2)p(n_1, n_2) = \alpha p(n_1 - 1, n_2) + \beta_1 p(n_1 + 1, n_2 - 1)$$
$$+ \beta_2 p(n_1, n_2 + 1); \qquad n_1, n_2 > 0$$

Verify that $p(n_1, n_2) = (1 - \rho_1)(1 - \rho_2)\rho_1^{n_1}\rho_2^{n_2}$ *where* $\max(\rho_1, \rho_2) < 1$ *and* $\rho_1 = \alpha/\beta_1$ *and* $\rho_2 = \alpha/\beta_2$.

2 Find the steady-state distribution (assumed to exist) for the network of M/M/1 systems given in Fig. 9.6. Write down the conditions which must be satisfied for the steady state to exist.
Write down expressions for the average time in the system for:

(a) a customer who takes the route (1, 2, 4);
(b) a customer who takes the route (3, 4).

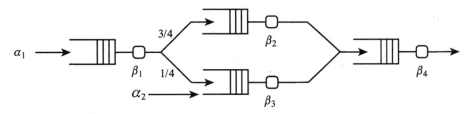

Fig. 9.6

9.9   Solutions to Examples 17

1 Let $p(n_1, n_2; t) = \Pr(N_1(t) = n_1, N_2(t) = n_2)$. Then to first order in δt:

$$p(0, 0; t + \delta t) = p(0, 0; t)(1 - \alpha\delta t) + p(0, 1; t)\beta_2\delta t$$
$$p(n_1, 0; t + \delta t) = p(n_1 - 1, 0; t)\alpha\delta t + p(n_1, 1; t)\beta_2\delta t$$
$$+ (1 - \alpha\delta t)(1 - \beta_1\delta t)p(n_1, 0; t) \qquad \text{for } n_1 > 0$$
$$p(0, n_2; t + \delta t) = p(1, n_2 - 1; t)\beta_1\delta t + p(0, n_2 + 1; t)\beta_2\delta t$$
$$+ (1 - \alpha\delta t)(1 - \beta_2\delta t)p(0, n_2; t) \qquad \text{for } n_2 > 0$$
$$p(n_1, n_2; t + \delta t) = p(n_1 - 1, n_2; t)\alpha\delta t$$
$$+ p(n_1 + 1, n_2 - 1; t)\beta_1\delta t + p(n_1, n_2 + 1; t)\beta_2\delta t$$
$$+ (1 - \alpha\delta t)(1 - \beta_1\delta t)(1 - \beta_2\delta t)p(n_1, n_2; t) \qquad \text{for } n_1, n_2 > 0$$

If we rearrange these equations and let $\delta t \rightarrow 0$ we obtain

$$\frac{dp(0,0;t)}{dt} = -\alpha p(0,0;t) + \beta_2 p(0,1;t)$$

$$\frac{dp(n_1,0;t)}{dt} = \alpha p(n_1 - 1,0;t) + \beta_2 p(n_1.1;t) - (\alpha + \beta_1)p(n_1,0;t)$$

$$\frac{dp(0,n_2;t)}{dt} = \beta_1 p(1,n_2 - 1;t) + \beta_2 p(0,n_2 + 1;t) - (\alpha + \beta_2)p(0,n_2;t)$$

$$\frac{dp(n_1,n_2;t)}{dt} = \alpha p(n_1 - 1,n_2;t) + \beta_1 p(n_1 + 1, n_2 - 1;t) + \beta_2 p(n_1,n_2 + 1;t)$$

$$-(\alpha + \beta_1 + \beta_2)p(n_1,n_2;t)$$

As $t \rightarrow \infty$, provided a steady-state solution exists, the derivatives on the L.H.S. approach zero and $p(n_1, n_2; t) \rightarrow p(n_1, n_2)$. Thus

$$\alpha p(0, 0) = \beta_2 p(0, 1)$$
$$(\alpha + \beta_1)p(n_1, 0) = \alpha p(n_1 - 1, 0) + \beta_2 p(n_1, 1)$$
$$(\alpha + \beta_2)p(0, n_2) = \beta_1 p(1, n_2 - 1) + \beta_2 p(0, n_2 + 1)$$
$$(\alpha + \beta_1 + \beta_2)p(n_1, n_2) = \alpha p(n_1 - 1, n_2) + \beta_1 p(n_1 + 1, n_2 - 1) + \beta_2 p(n_1, n_2 + 1)$$

If we assume the result $p(n_1, n_2) = (1 - \rho_1)(1 - \rho_2)\rho_1^{n_1}\rho_2^{n_2}$ the L.H.S. of the first equation becomes $\alpha(1 - \rho_1)(1 - \rho_2)$. The R.H.S. of the first equation is

$$\beta_2(1 - \rho_1)(1 - \rho_2)\frac{\alpha}{\beta_2} = \alpha(1 - \rho_1)(1 - \rho_2)$$

as required. For the last equation, the L.H.S. is

$$\alpha(1 - \rho_1)(1 - \rho_2)\left(\frac{\alpha}{\beta_1}\right)^{n_1}\left(\frac{\alpha}{\beta_2}\right)^{n_2} + \alpha(1 - \rho_1)(1 - \rho_2)\left(\frac{\alpha}{\beta_1}\right)^{n_1 - 1}\left(\frac{\alpha}{\beta_2}\right)^{n_2}$$

$$+ \alpha(1 - \rho_1)(1 - \rho_2)\left(\frac{\alpha}{\beta_1}\right)^{n_1}\left(\frac{\alpha}{\beta_2}\right)^{n_2 - 1}$$

The R.H.S. is

$$\alpha(1 - \rho_1)(1 - \rho_2)\left(\frac{\alpha}{\beta_1}\right)^{n_1 - 1}\left(\frac{\alpha}{\beta_2}\right)^{n_2} + \beta_1 \frac{\alpha}{\beta_1}(1 - \rho_1)(1 - \rho_2)\left(\frac{\alpha}{\beta_1}\right)^{n_1}\left(\frac{\alpha}{\beta_2}\right)^{n_2 - 1}$$

$$+ \beta_2 \frac{\alpha}{\beta_2}(1 - \rho_1)(1 - \rho_2)\left(\frac{\alpha}{\beta_1}\right)^{n_1}\left(\frac{\alpha}{\beta_2}\right)^{n_2}$$

It is clear that we have equality. Verification of the other two equations is equally straightforward.

2 In an obvious notation

$$p(n_1, n_2, n_3, n_4) = \Pr(N_1 = n_1, N_2 = n_2, N_3 = n_3, N_4 = n_4)$$
$$= (1 - \rho_1)(1 - \rho_2)(1 - \rho_3)(1 - \rho_4)\rho_1^{n_1}\rho_2^{n_2}\rho_3^{n_3}\rho_4^{n_4}$$

where $\rho_1 = \alpha_1/\beta_1$, $\rho_2 = (\frac{3}{4}\alpha_1)/\beta_2$, $\rho_3 = (\frac{1}{4}\alpha_1 + \alpha_2)/\beta_3$, $\rho_4 = (\alpha_1 + \alpha_2)/\beta_4$ and $\max(\rho_1, \rho_2, \rho_3, \rho_4) < 1$.

(a) In an obvious notation the expected time in the system is

$$E[W_1] + \frac{1}{\beta_1} + E[W_2] + \frac{1}{\beta_2} + E[W_4] + \frac{1}{\beta_4}$$

$$= \frac{\alpha_1}{\beta_1(\beta_1 - \alpha_1)} + \frac{1}{\beta_1} + \frac{\frac{3}{4}\alpha_1}{\beta_2\left(\beta_2 - \frac{3}{4}\alpha_1\right)} + \frac{1}{\beta_2} + \frac{\alpha_1 + \alpha_2}{\beta_4(\beta_4 - \alpha_1 - \alpha_2)} + \frac{1}{\beta_4}$$

where we have used (2.13) for the expected time in the queue.

(b) In the same way the expected time in the system is

$$E[W_3] + \frac{1}{\beta_3} + E[W_4] + \frac{1}{\beta_4}$$

$$= \frac{\alpha_2 + \frac{1}{4}\alpha_1}{\beta_3\left(\beta_3 - \alpha_2 - \frac{1}{4}\alpha_1\right)} + \frac{1}{\beta_3} + \frac{\alpha_1 + \alpha_2}{\beta_4(\beta_4 - \alpha_1 - \alpha_2)} + \frac{1}{\beta_4}$$

9.10   Open networks

The product-form type steady-state solution which was obtained for feed-forward networks also applies to an open network of systems in which a customer can visit a system more than once.

We consider a network of M systems (each of the M/M/1 type). Arrivals from the 'outside world' occur at system i at random at rate λ_i. Service at system i has a negative exponential distribution at rate β_i. A departure from system i joins system j with probability P_{ij} or returns to the 'outside world' with probability $1 - \sum_{j=1}^{M} P_{ij}$ ($= \gamma_i$, say). Thus the actual arrivals to system i comprise those from outside plus those who join it from within the network (this could include feedback from its own departures with probability P_{ii}). Thus the input to system i ($i = 1, 2, ..., M$) is not necessarily a Poisson stream. If α_i is the total arrival rate to system i, then

$$\alpha_i = \lambda_i + \sum_{j=1}^{M} \alpha_j P_{ji} \qquad \text{for } i = 1, 2, ..., M \qquad (9.14)$$

i.e. in matrix form $\boldsymbol{a} = \lambda(\mathbf{I} - \mathbf{P})^{-1}$ provided $\mathbf{I} - \mathbf{P}$ is not singular. The last requirement means that every incoming customer eventually leaves the network with probability one. An example of an open network (with $M = 3$) is shown in Fig. 9.7. The result we shall prove is that if $p(n_1, n_2, \ldots, n_M)$ is the steady-state probability of n_i customers in system i then

$$p(n_1, n_2, \ldots, n_M) = \prod_{i=1}^{M} (1 - \rho_i)\rho_i^{n_i} \qquad \text{where } \rho_i = \frac{\alpha_i}{\beta_i} \qquad (9.15)$$

i.e. each system behaves like an independent M/M/1 system with arrival rate α_i and service rate β_i.

We shall leave this as an exercise (question 1 of Examples 18). We shall establish an even more general result. We assume that the external arrivals all come from a single 'source' S and that a proportion P_{si} of these arrivals go to system i. Furthermore, we assume that the total arrival rate from S is $\lambda(n)$ where $n = (n_1 + n_2 + \cdots + n_M)$ is the number of customers in the network. We assume that customers who leave the network altogether from system i go to a 'terminus' T with probability P_{iT} ($=\gamma_i$). There can be no feedback into the source or from the terminus. We further assume that the service rate at system i is $\beta_i(n_i)$ and depends on the number in system i. We leave it as an exercise to show that our original idea of an open network is a special case of this more general definition which is shown in Fig. 9.8. It is convenient to use a vector notation and let

$$p(\boldsymbol{n}; t) = p(N_1(t) = n_1, \ldots, N_M(t) = n_M)$$

where n_i means n_i customers in system i at time t and $\boldsymbol{n} = (n_1, n_2, \ldots, n_M)$. For the steady state, $p(\boldsymbol{n}) = \text{Limit}_{t\to\infty} \, p(\boldsymbol{n}; t)$. It is convenient to let

$$\mathbf{1}_i = (0, 0, \ldots, \underset{i\text{th column}}{1}, 0, 0)$$

Using the same type of argument as in Chapter 3 for birth–death models, we relate the situation at time $t + \delta t$ to that at time t.

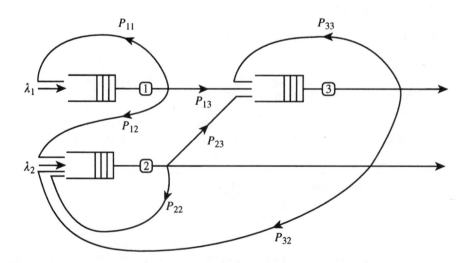

Fig. 9.7

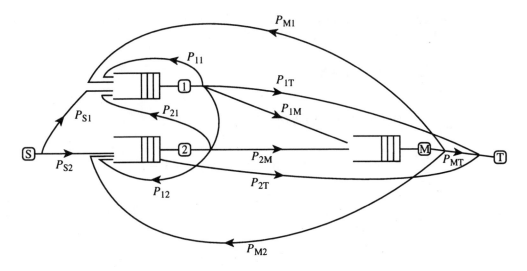

Fig. 9.8

Then we let $t \to \infty$ and obtain for the steady-state probabilities:

$$\lambda(n)p(\boldsymbol{n}) + \sum_{i=1}^{M} \beta_i(n_i)p(\boldsymbol{n}) = \lambda(n-1) \sum_{i=1}^{M} P_{si}p(\boldsymbol{n} - \mathbf{1}_i)$$

$$+ \sum_{i=1}^{M} P_{iT}\beta_i(n_i + 1)p(\boldsymbol{n} + \mathbf{1}_i)$$

$$+ \sum_{i=1}^{M}\sum_{j=1}^{M} P_{ji}\beta_j(n_j + 1 - \delta_{ij})p(\boldsymbol{n} + \mathbf{1}_j - \mathbf{1}_i) \qquad (9.16)$$

which can be put in the form

$$\lambda(n)p(\boldsymbol{n}) - \sum_{i=1}^{M} P_{iT}\beta_i(n_i + 1)p(\boldsymbol{n} + \mathbf{1}_i) = \lambda(n-1) \sum_{i=1}^{M} P_{si}p(\boldsymbol{n} - \mathbf{1}_i)$$

$$- \sum_{i=1}^{M} \beta_i(n_i)p(\boldsymbol{n}) + \sum_{i=1}^{M}\sum_{j=1}^{M} P_{ji}\beta_j(n_j + 1 - \delta_{ij})p(\boldsymbol{n} + \mathbf{1}_j - \mathbf{1}_i)$$

We also have

$$\lambda(0)p(\boldsymbol{0}) - \sum_{i=1}^{M} P_{iT}\beta_i(1)p(\mathbf{1}_i) = 0$$

It is convenient to introduce the quantities e_i (related to α_i?) defined by

$$e_i = P_{si} + \sum_{j=1}^{M} e_j P_{ji} \text{ so that } P_{si} = e_i - \sum_{j=1}^{M} e_j P_{ji} \qquad \text{for } i = 1, \dots, M$$

e_i is the expected number of times a customer visits system i whilst in the network. If we substitute for P_{si} we obtain, after a little rearrangement,

$$\lambda(n)p(n) - \sum_{i=1}^{M} P_{iT}\beta_i(n_i + 1)p(n + 1_i)$$

$$= \sum_{i=1}^{M}\left[\lambda(n-1)e_ip(n-1_i) - \beta_i(n_i)p(n)\right.$$

$$\left. - \sum_{j=1}^{M} P_{ji}\{\lambda(n-1)e_jp(n-1_i) - \beta_j(n_j+1-\delta_{ij})p(n+1_j-1_i)\}\right]$$

$$\therefore\ \lambda(n)p(n) - \sum_{i=1}^{M} P_{iT}\beta_i(n_i + 1)p(n + 1_i) = \sum_{i=1}^{M}\left[\lambda(n-1)e_ip(n-1_i) - \beta_i(n_i)p(n)\right.$$

$$\left. - \sum_{j=1}^{M} P_{ji}\{\lambda(n-1)e_jp(n+1_j-1_i-1_j) - \beta_j(n_j+1-\delta_{ij})p(n+1_j-1_i)\}\right] \quad (9.17)$$

The form of this equation is

$$C(n) = \sum_{i=1}^{M}\left[D_i(n) - \sum_{j=1}^{M} P_{ji}D_j(n + 1_j - 1_i)\right] \quad (9.18)$$

where

$$C(n) = \lambda(n)p(n) - \sum_{i=1}^{M} P_{iT}\beta_i(n_i + 1)p(n + 1_i)$$

and

$$D_i(n) = \lambda(n-1)e_ip(n-1_i) - \beta_i(n_i)p(n)$$

A solution is obtained with $D_i(n) = 0$ for all i and n when also $C(n) = 0$ for all n as it should be if the equations are to hold.

This latter condition follows for if $D_i(n + 1_i) = 0$ for all n and i

$$\lambda(n)e_ip(n) - \beta_i(n_i + 1)p(n + 1_i) = 0 \qquad \text{for all } i$$

Multiply this by P_{iT} and sum over $i = 1, 2, \ldots, M$. Thus

$$\lambda(n)\sum_{i=1}^{M} e_iP_{iT}p(n) - \sum_{i=1}^{M} P_{iT}\beta_i(n_i + 1)p(n + 1_i) = 0$$

$$\therefore\ \lambda(n)\sum_{i=1}^{M} e_i\left(1 - \sum_{j=1}^{M} P_{ij}\right)p(n) - \sum_{i=1}^{M} P_{iT}\beta_i(n_i + 1)p(n + 1_i) = 0$$

$$\therefore\ \lambda(n)\left[\sum_{i=1}^{M} e_i - \sum_{i=1}^{M}\sum_{j=1}^{M} e_jP_{ji}\right]p(n) - \sum_{i=1}^{M} P_{iT}\beta_i(n_i + 1)p(n + 1_i) = 0$$

[N.B. Changing the suffix labels in the summation is valid.]

$$\therefore \ \lambda(n)p(n) \sum_{i=1}^{M} P_{si} - \sum_{i=1}^{M} P_{iT}\beta_i(n_i + 1)p(n + 1_i) = 0$$

$$\therefore \ \lambda(n)p(n) - \sum_{i=1}^{M} P_{iT}\beta_i(n_i + 1)p(n + 1_i) = 0$$

i.e.

$$C(n) = 0$$

Thus $D_i(n) = 0$ for all i and n ensures that all the equations are satisfied. $D_i(n) = 0$ gives $p(n) = \lambda(n-1)e_i p(n-1_i)/\beta(n_i)$ and is the vector equivalent of $p_n = (a_{n-1})p_{n-1}/\beta_n$ (3.3). Thus we obtain

$$p(n) = p(0) \prod_{k=1}^{n} \lambda(k-1) \prod_{i=1}^{M} \frac{e_i^{n_i}}{Q_i(n_i)} \tag{9.19}$$

where $Q_i(n_i) = \prod_{k=1}^{n_i} \beta_i(k)$. Since the process we are dealing with is a Markov process this solution is in fact the unique steady-state solution. Of course $\lambda(n)e_i$ is the total arrival rate to system i when there are n in the network. Of these $\lambda(n)P_{si}$ are the external arrivals, the rest $\lambda(n) \sum_{j=1}^{M} e_j P_{ji}$ are transfers (feedback) from within the network.

In the case where $\lambda(n) = \lambda$ is constant for all n, then $\lambda(n)e_i$ is equivalent to a_i, the total arrival rate to system i (as at (9.14)). Then

$$p(n) = \prod_{i=1}^{M} p_i(n_i) \tag{9.20}$$

where

$$p_i(n_i) = \frac{(\lambda e_i)^{n_i}}{\beta_i(1)\beta_i(2) \dots \beta_i(n_i)} \ p_i(0)$$

is the marginal distribution of n_i.

$$p_i(0) = 1 \Big/ \left\{ \sum_{k=0}^{\infty} \frac{(\lambda e_i)^k}{\beta_i(1)\beta_i(2) \dots \beta_i(k)} \right\} \tag{9.21}$$

is analogous with the result (3.4).

The argument leading to (9.20) and (9.21) is not easy. The logic of the modelling is analogous to the derivation of the birth–death equations in Chapter 3 and the solution of question 1 of Examples 17. The latter is a relatively easy special case. The algebra involved in the derivation of (9.20) and (9.21) makes demands on our algebraic skills. The reader is advised to carefully work through all the steps required.

Example 9.3

The network illustrated in Fig. 9.9 represents the production process for certain articles made in a factory. Processing times are assumed to be exponentially distributed. 'Skeleton' parts arrive at random at station 1 at an average rate α and are processed at

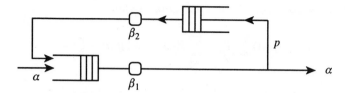

Fig. 9.9

rate β_1. A proportion p of this output is faulty and has to be 'partially dismantled' at station 2 at rate β_2 before being returned to station 1.

Find the mean number of articles in each of the sub-systems at stations 1 and 2. Find the mean time that an article spends in the system.

The net arrival rates at the two stations are $\lambda_1 = \alpha + \lambda_2$, and $\lambda_2 = p\lambda_1$

$$\therefore \ \lambda_1 = \frac{\alpha}{1-p}, \qquad \lambda_2 = \frac{p\alpha}{1-p}$$

Each system behaves like an M/M/1 system. Thus

$$E[N_1] = \frac{\rho_1}{1-\rho_1} \text{ where } \rho_1 = \frac{\alpha}{\beta_1(1-p)}$$

$$E[N_2] = \frac{\rho_2}{1-\rho_2} \text{ where } \rho_2 = \frac{p\alpha}{\beta_2(1-p)}$$

and both ρ_1 and ρ_2 are less than 1. If the time spent in the system is T, then we need to calculate $W = E[T]$. Now for a given 'skeleton' article which spends time t in the system, the mean number of new arrivals in the system during its time in the system is

$$E[N_1 + N_2 \,|\, T = t] = \alpha t$$

This is because we have random arrivals at rate α during this time and these articles which are still in the system at the moment of departure of the given 'skeleton' article are either at station 1 or station 2. Thus

$$E[N_1 + N_2] = \alpha E[T] = \alpha W$$

Compare this result with (2.15) and (4.7) and the remark following (4.7).

$$\therefore \ W = \frac{1}{\alpha} \left[\frac{\rho_1}{1-\rho_1} + \frac{\rho_2}{1-\rho_2} \right]$$

This argument is undoubtedly the simplest way to obtain an expression for W. We can obtain the same result by somewhat more involved reasoning. Let W_1 and W_2 be the average times spent in the system of station 1 and station 2, respectively, for an article that is processed at the respective stations. Then from (2.15) or indeed (4.7) for an M/M/1 system

$$W_1 = \frac{E[N_1]}{\lambda_1} = \frac{(1-p)E[N_1]}{\alpha}$$

and

$$W_2 = \frac{E[N_2]}{\lambda_2} = \frac{(1-p)E[N_2]}{p\alpha}$$

Now an article makes one 'pass' through the system with probability $(1 - p)$, and spends mean time W_1 in the system.

An article makes two passes through the system with probability $p(1 - p)$, and in this case spends mean time $W_1 + W_2$ in the system.

An article makes three passes through the system with probability $p^2(1 - p)$, and in this case spends mean time $3W_1 + 2W_2$ in the system, etc.

These probabilities are of course the proportion of incoming articles which make one pass, two passes, three passes, etc. Thus the mean time spent in the system by an incoming article is

$$W = (1 - p)W_1 + p(1 - p)(2W_1 + W_2) + p^2(1 - p)(3W_1 + 2W_2) + \cdots$$

$$\therefore\ W = (1 - p)W_1[1 + 2p + 3p^2 + 4p^3 + \cdots] + p(1 - p)W_2[1 + 2p + 3p^2 + \cdots]$$

$$= \frac{W_1}{1 - p} + \frac{pW_2}{1 - p}$$

[since $1 + 2p + 3p^2 + \cdots = 1/(1 - p)^2$; see derivation of (2.10)]

$$\therefore\ W = \frac{E[N_1] + E[N_2]}{\alpha}.$$

as before.

9.11 Examples 18

1 Show that the first example of an open network as given in the initial discussion on open networks is a special case of the more general system described a little later.

   Identify $\lambda(n)$ $(=\lambda)$ in terms of $\lambda_1, \lambda_2, \ldots, \lambda_M$ and write down the value of P_{si} $(i = 1, 2, \ldots, M)$. Identify α_i and γ_i.

   Write down the steady-state equations (not in vector form) for this special network of M/M/1 systems and obtain the solution in the form corresponding to (9.15).

2 The network shown in Fig. 9.10 illustrates the flow of work to a machine (1) in a factory. Jobs arrive at random from another part of the factory at rate λ. Processing time on machine 1 is exponential at rate β_1.

   A proportion p of the output jobs are faulty and these jobs have to be partly 'unprocessed' at machine 2 where service time is exponential at rate β_2. They are then reprocessed by machine 1.

   Find the expected number of jobs in each system of the network and the average time that a job spends in the network.

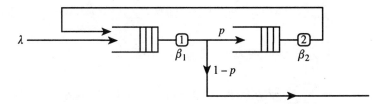

Fig. 9.10

9.12   Solutions to Examples 18

1 $\lambda(n) = \lambda_1 + \lambda_2 + \cdots + \lambda_M$
$P_{si} = \lambda_i / (\lambda_1 + \lambda_2 + \cdots + \lambda_M)$
so that $\lambda(n)P_{si} = \lambda_i$.

$$\gamma_i = 1 - \sum_{j=1}^{M} P_{ij} = P_{iT}$$

$$\alpha_i = \lambda(n)e_i$$

and the equation

$$e_i = P_{si} + \sum_{j=1}^{M} e_j P_{ji}$$

becomes, after multiplication by $\lambda(n)$,

$$\alpha_i = \lambda_i + \sum_{j=1}^{M} \alpha_j P_{ji}$$

Since e_i represents the expected number of visits a customer makes to system i then, provided this is finite (so α_i is finite), all customers eventually leave the network. They do not visit any system infinitely often.

If we consider the transitions that can occur in $(t, t + \delta t)$ (these are arrivals from outside, departures to the outside, transfers from system j to system i after service completion at system j):

$$p(n_1, n_2, \ldots, n_M; t + \delta t) = \sum_{i=1}^{M} p(n_1, n_2, \ldots, n_i - 1, \ldots, n_M; t)\lambda_i \delta t$$

$$+ \sum_{i=1}^{M} p(n_1, n_2, \ldots, n_M)(1 - \lambda_i \delta t)(1 - \beta_i \delta t)$$

$$+ \sum_{j=1}^{M} p(n_1, n_2, \ldots, n_j + 1, \ldots, n_M)\beta_j \gamma_j \delta t$$

$$+ \sum_{i=1}^{M} \sum_{j=1}^{M} p(n_1, n_2, \ldots, n_i - 1, \ldots, n_j + 1, \ldots, n_M)\beta_j P_{ji} \delta t$$

Then

$$\frac{dp}{dt}(n_1, n_2, \ldots, n_M; t)$$

$$= \sum_{i=1}^{M} \lambda_i p(n_1, n_2, \ldots, n_i - 1, \ldots, n_M; t) - \sum_{i=1}^{M} (\lambda_i + \beta_i)p(n_1, n_2, \ldots, n_M; t)$$

$$+ \sum_{i=1}^{M} \beta_i \gamma_i p(n_1, n_2, \ldots, n_i + 1, \ldots, n_M; t)$$

$$+ \sum_{i=1}^{M} \sum_{j=1}^{M} \beta_j P_{ji} p(n_1, n_2, \ldots, n_i - 1, \ldots, n_j + 1, \ldots, n_M; t)$$

Thus in the steady state in which the t dependence on the R.H.S. disappears and the derivative on the L.H.S. becomes zero

$$\sum_{i=1}^{M} \lambda_i p(n_1, n_2, \ldots, n_M) - \sum_{i=1}^{M} \beta_i \gamma_i p(n_1, n_2, \ldots, n_i + 1, \ldots, n_M)$$

$$= \sum_{i=1}^{M} \lambda_i p(n_1, n_2, \ldots, n_i - 1, \ldots, n_M) - \sum_{i=1}^{M} \beta_i p(n_1, n_2, \ldots, n_M)$$

$$+ \sum_{i=1}^{M} \sum_{j=1}^{M} \beta_j P_{ji} p(n_1, n_2, \ldots, n_i - 1, \ldots, n_j + 1, \ldots, n_M)$$

Now $\lambda_i = \alpha_i - \Sigma_{j=1}^{M} \alpha_j P_{ji}$

$$\therefore \sum_{i=1}^{M} \lambda_i p(n_1, n_2, \ldots, n_M) - \sum_{i=1}^{M} \beta_i \gamma_i p(n_1, n_2, \ldots, n_i + 1, \ldots, n_M)$$

$$= \sum_{i=1}^{M} \left[\alpha_i p(n_1, n_2, \ldots, n_i - 1, \ldots, n_M) - \beta_i p(n_1, n_2, \ldots, n_M) \right.$$

$$- \sum_{j=1}^{M} P_{ji} \{ \alpha_j p(n_1, n_2, \ldots, n_i - 1, \ldots, n_j, \ldots, n_M)$$

$$\left. - \beta_j p(n_1, n_2, \ldots, n_i - 1, \ldots, n_j + 1, \ldots, n_M) \} \right]$$

Now these equations take the form

$$C(n_1, n_2, \ldots, n_M)$$

$$= \sum_{i=1}^{m} \left[D_i(n_1, \ldots, n_M) - \sum_{j=1}^{M} P_{ji} D_j(n_1, n_2, n_i - 1, \ldots, n_j + 1, \ldots, n_M) \right]$$

where

$$C(n_1, n_2, \ldots, n_M)$$

$$= \sum_{i=1}^{M} \lambda_i p(n_1, n_2, \ldots, n_M) - \sum_{i=1}^{M} \beta_i \gamma_i p(n_1, n_2, \ldots, n_i + 1, \ldots, n_M)$$

and

$$D_i(n_1, \ldots, n_M) = \alpha_i p(n_1, n_2, \ldots, n_i - 1, \ldots, n_M)$$
$$- \beta_i p(n_1, n_2, \ldots, n_i, \ldots, n_M)$$

Now if $D_i(n_1, \ldots, n_M) = 0$ for all i and $(n_1, n_2, \ldots, n_M)$ then

$$\alpha_i p(n_1, n_2, \ldots, n_i, \ldots, n_M) - \beta_i p(n_1, n_2, \ldots, n_i + 1, \ldots, n_M) = 0 \qquad \text{for all } i$$

Multiply by γ_i and sum over $i = 1, 2, ..., M$.

$$\therefore \sum_{i=1}^{M} \alpha_i \gamma_i p(n_1, n_2, ..., n_M) - \sum_{i=1}^{M} \beta_i \gamma_i p(n_1, n_2, ...n_i + 1, ..., n_M) = 0$$

$$\therefore \sum_{i=1}^{M} \alpha_i \left(1 - \sum_{j=1}^{M} P_{ij}\right) p(n_1, n_2, ..., n_M) - \sum_{i=1}^{M} \beta_i \gamma_i p(n_1, n_2, ..., n_i + 1, ..., n_M) = 0$$

$$\therefore \sum_{i=1}^{M} \left[\alpha_i - \sum_{j=1}^{M} \alpha_j P_{ji}\right] p(n_1, n_2, ..., n_M) - \sum_{i=1}^{M} \beta_i \gamma_i p(n_1, n_2, ..., n_{i+1}, ..., n_M) = 0$$

$$\therefore \sum_{i=1}^{M} \lambda_i p(n_1, n_2, ..., n_M) - \sum_{i=1}^{M} \beta_i \gamma_i p(n_1, n_2, ..., n_i + 1, ..., n_M) = 0$$

i.e.

$$C(n_1, n_2, ..., n_M) = 0$$

Thus $D_i(n_1, n_2, ..., n_M) = 0$ for all i and $(n_1, n_2, ..., n_M)$ ensures that all the equations are satisfied. But $D_i(n_1, n_2, ..., n_M) = 0$ means

$$p(n_1, n_2, ..., n_M) = \frac{\alpha_i}{\beta_i} p(n_1, n_2, ..., n_i - 1, ..., n_M)$$

Thus

$$p(n_1, n_2, ..., n_M) = p(0, 0, ..., 0) \prod_{i=1}^{M} \left(\frac{\alpha_i}{\beta_i}\right)^{n_i}$$

and since $\sum p(n_1, n_2, ..., n_M) = 1$ when summed over all $(n_1, n_2, ..., n_M)$,

$$p(0, 0, ..., 0) = \prod_{i=1}^{M} \left(1 - \frac{\alpha_i}{\beta_i}\right)$$

$$\therefore p(n_1, n_2, ..., n_M) = \prod_{i=1}^{M} \left(1 - \frac{\alpha_i}{\beta_i}\right)\left(\frac{\alpha_i}{\beta_i}\right)^{n_i}$$

2 The equations which determine the arrival rates α_1, α_2 to the two systems are $\alpha_1 = \lambda + \alpha_2$ and $\alpha_2 = p\alpha_1$

$$\therefore \quad \alpha_1 = \frac{\lambda}{1 - p}, \qquad \alpha_2 = \frac{p\lambda}{1 - p}$$

Each system behaves like an M/M/1 system so that

$$E[N_1] = \frac{\rho_1}{1 - \rho_1} \quad \text{and} \quad E[N_2] = \frac{\rho_2}{1 - \rho_2}$$

where $\rho_1 = \alpha_1/\beta_1$, $\rho_2 = \alpha_2/\beta_2$.

The mean time spent in the system is $E[T]$ where $\lambda E[T] = E[N_1 + N_2]$. This is the equivalent of $L = \lambda E[T]$ $(= \alpha W)$ as at (2.15), (3.12) and (4.7).

$$\therefore \; E[T] = \frac{1}{\lambda} \left(\frac{\rho_1}{1-\rho_1} + \frac{\rho_2}{1-\rho_2} \right) = \frac{1}{\beta_1(1-p)-\lambda} + p\,\frac{1}{\beta_2(1-p)-p\lambda}$$

$$= \frac{1}{1-p}\,\frac{1}{\beta_1-\alpha_1} + \frac{p}{1-p}\,\frac{1}{\beta_2-\alpha_2}$$

$$= \frac{1}{1-p}\,[E[T_1] + pE[T_2]]$$

where $E[T_1]$ and $E[T_2]$ are the respective expected times in the system for the corresponding M/M/1 systems.

Notice that the number of times a job visits the first system is a random variable R, where

$$\Pr(R = r) = p^{r-1}(1-p); \qquad r = 1, 2, \ldots$$

Thus

$$E[R] = (1-p) \sum_{r=0}^{\infty} rp^{r-1} = \frac{1}{1-p}$$

A proportion p of jobs that visit system 1 also visit system 2, hence the result. Our analysis above obtains the result more or less automatically without having to worry about the number of visits. The result above can be put in the form

$$E[T] = E[R]E[T_1] + pE[R]E[T_2]$$

9.13 Closed networks

We can extend the product-form result on open networks to closed networks. In the latter there are no external arrivals and a fixed number n, say, of customers circulate between the systems which make up the network.

If $p(n) = \Pr(N_1 = n_1, N_2 = n_2, \ldots, N_M = n_M)$ is the steady-state probability that there are n_i customers in system i, for $i = 1, \ldots, M$, then since

$$n_1 + n_2 + \cdots + n_M = n \tag{9.22}$$

we notice that the variables N_i are not independent, as was the case in open networks. Since there are no external arrivals or departures, the equations for $p(n)$ take the form

$$\sum_{i=1}^{M} \sum_{j=1}^{M} P_{ji}\beta_j(n_j + 1 - \delta_{ij})p(n + 1_j - 1_i) - \sum_{i=1}^{M} \beta_i(n_i)p(n) = 0 \tag{9.23}$$

Since $P_{si} = 0$ the equations for the e_i now take the form

$$e_i = \sum_{j=1}^{M} e_j P_{ji} \quad \text{or} \quad 1 = \sum_{j=1}^{M} \frac{e_j P_{ji}}{e_i}$$

Thus on using this (heavily disguised) value of 1 in (9.23) we obtain

$$\sum_{i=1}^{M}\sum_{j=1}^{M} P_{ji}\beta_j(n_j+1-\delta_{ij})p(n+1_j-1_i) - \sum_{i=1}^{M}\sum_{j=1}^{M} \frac{e_j P_{ji}}{e_i}\beta_i(n_i)p(n) = 0$$

i.e.

$$\sum_{i=1}^{M}\left[\sum_{j=1}^{M} P_{ji}\left\{\beta_j(n_j+1-\delta_{ij})p(n+1_j-1_i) - \frac{e_j}{e_i}\beta(n_i)p(n)\right\}\right] = 0$$

Thus it is clear that if the contents of the curly brackets are zero for all n and all values of $i, j = 1, 2, \ldots, M$ then our equations will hold. Thus a solution is obtained with

$$p(n) = \frac{e_i}{e_j}\frac{\beta_j(n_j+1-\delta_{ij})}{\beta(n_i)} \cdot p(n+1_j-1_i) \qquad \text{for all } n \text{ and } i, j$$

It is clear that the solution to this recurrence relationship is that

$$p(n) = \kappa(n,M)\prod_{i=1}^{M}\frac{e_i^{n_i}}{\beta_i(1)\beta_i(2)\ldots\beta_i(n_i)} \tag{9.24}$$

where κ is a constant that will depend upon n, the total number of customers in the closed network, and M, and is determined from the condition

$$\sum_n p(n) = 1 \tag{9.25}$$

The calculation of κ can present very serious problems for a large network (M large) containing a large number of customers (n large).

The e_i are such that $e = eP$ so that if π is the equilibrium distribution for the Markov chain with one-step transition matrix P, where $\pi = \pi P$ and $\pi_1 + \pi_2 + \cdots + \pi_M = 1$, we may take $e \propto \pi$ or indeed $e = \pi$. If $\alpha(n)$ denotes the sum of the arrival rates $\alpha_1 + \alpha_2 + \cdots + \alpha_M$ to all M systems when there are n customers in the network it makes sense to use $\alpha(n)$ as the constant of proportionality so that $e_i = \alpha(n)\pi_i = \alpha_i$, say. Then

$$p(n_1, n_2, \ldots, n_M) = \kappa(n,M)\prod_{i=1}^{M} p_i(n_i) \tag{9.26}$$

where

$$p_i(n_i) = \frac{\alpha_i^{n_i}}{\beta_i(1)\beta_i(2)\ldots\beta_i(n_i)} \tag{9.27}$$

in accord with the result (3.3).

Note that we have a product-form solution for the network, but the N_i, the numbers in each system, are NOT independent. The marginal distribution of N_i is NOT $p_i(n_i)$. This is so for the open networks described earlier.

Example 9.4

Figure 9.11 shows a model of a computer attached to a terminal. Programs are executed by the computer at rate β_1 and execution times have a negative exponential distribution.

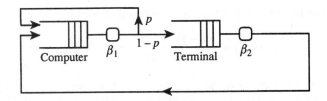

Fig. 9.11

A proportion p of these programs is immediately resubmitted. The rest need modification before resubmission from the terminal. The time to modify a program has a negative exponential distribution with mean $1/\beta_2$. If there are a total of n programs in the network, find the probability that r of them are in the computer system and $n-r$ of then are being modified or waiting at the terminal.

The equations for π are, since

$$\mathbf{P} = \begin{bmatrix} p & 1-p \\ 1 & 0 \end{bmatrix}$$

$$\pi_1 = p\pi_1 + \pi_2 \quad \text{and} \quad \pi_1 + \pi_2 = 1$$
$$\pi_2 = (1-p)\pi_1$$

Thus

$$\pi_1 = \frac{1}{2-p} \quad \text{and} \quad \pi_2 = \frac{1-p}{2-p}$$

Thus the arrival rates at the two systems are

$$a_1 = \frac{\alpha(n)}{2-p} \quad \text{and} \quad a_2 = \frac{\alpha(n)(1-p)}{2-p}$$

Thus with $\rho_1 = a_1/\beta_1$, $\rho_2 = a_2/\beta_2$

$$p(n_1, n_2) = p(n_1, n-n_1) = \kappa(1-\rho_1)\rho_1^{n_1}(1-\rho_2)\rho_2^{n-n_1}, \qquad 0 \leqslant n_1 \leqslant n$$

Since

$$\sum_{n_1, n_2} p(n_1, n_2) = 1, \qquad \kappa(1-\rho_1)(1-\rho_2)\sum_{n_1=0}^{n} \rho_1^{n_1}\rho_2^{n-n_1} = 1$$

$$\therefore \kappa(1-\rho_1)(1-\rho_2)\rho_2^{n}\sum_{n_1=0}^{n}\left(\frac{\rho_1}{\rho_2}\right)^{n_1} = 1$$

$$\therefore \kappa(1-\rho_1)(1-\rho_2)\rho_2^{n}\,\frac{1-\left(\dfrac{\rho_1}{\rho_2}\right)^{n+1}}{1-\dfrac{\rho_1}{\rho_2}} = 1$$

$$\therefore p(n_1, n-n_1) = \frac{1-\gamma}{1-\gamma^{n+1}}\,\gamma^{n_1}$$

where $\gamma = \rho_1/\rho_2 = \beta_2/((1-p)\beta_1)$ and we note that this does not involve $\alpha(n)$. Thus

$$p(r, n-r) = \frac{(1-\gamma)}{1-\gamma^{n+1}} \gamma^r, \qquad r = 0, 1, 2, \ldots, n$$

The proportion of time the computer is free is $p(0, n)$ and this is given by $(1-\gamma)/(1-\gamma^{n+1})$. We can find α_1 (and hence $\alpha(n)$) by equating this to $1 - \alpha_1/\beta_1$.

9.14 Examples 19

1 A closed queueing network has n customers in M systems. Show that the number of state probabilities $p(\boldsymbol{n})$ is

$$\binom{n + M - 1}{M - 1}$$

This shows the number of terms in the sum $\sum_n p(\boldsymbol{n}) = 1$, and illustrates in part the difficulty of evaluating $\kappa(n, M)$ in (9.24) and (9.26).

2 Consider the machine interference problem (first considered in Chapter 3) with one operator as a closed queueing network with N customers (the N machines). When a machine breaks down it has to be repaired by the operator and the repair time is negative exponential at rate β (Fig. 9.12). Upon completion of the repair the running machine joins the first system where service times (the running times) have a negative exponential distribution with mean $1/\alpha$. Thus when r machines are in the first system the 'repair' rate α_r is $r\alpha$. Find the probability that there are n machines stopped and $N - n$ running.

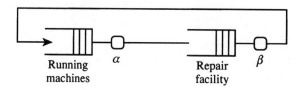

Running α Repair β
machines   facility

Fig. 9.12

3 The queueing network shown in Fig. 9.13 has as its customers components for a machine. The machine needs two working components to operate at full efficiency. The lifetime of a running component has a negative exponential distribution with mean $1/\beta_1$. When a component fails it goes to a repair system where it is repaired by an operator whose repair time has a negative exponential

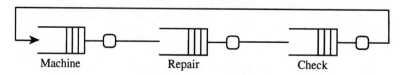

Machine   Repair   Check

Fig. 9.13

distribution with mean $1/\beta_2$. It is then checked by another operator, the checking time having a negative exponential distribution with mean $1/\beta_3$. It then joins the spare components and it is from these that a replacement is taken (provided one is available) when a component fails in the machine. There are a total of five components in the network. Let N_1 represent the number of working components and spares, N_2 the number under repair or waiting for repair, and N_3 the number being checked or waiting to be checked. Show how to find the proportion of time the machine is unable to operate at full efficiency.

9.15 Solutions to Examples 19

1 The number of states can be thought of as the number of non-negative integer solutions of the equation $n_1 + n_2 + \cdots + n_M = n$. With a cross ($\times$) to represent a customer this is the number of ways of putting n crosses and $M - 1$ strokes ($/$) in a line. Thus with $n = 7$ and $M = 3$

$$\times\times/\times\times\times\times/\times \text{ represents } n_1 = 2, n_2 = 4, n_3 = 1$$
$$/\times\times\times/\times\times\times\times \text{ represents } n_1 = 0, n_2 = 3, n_3 = 4, \text{ etc.}$$

The number of arrangements is

$$\binom{n + M - 1}{M - 1}$$

2 In this case $\pi_1 = \pi_2 = 1/2$ since

$$\mathbf{P} = \begin{pmatrix} 0 & 1 \\ 1 & 0 \end{pmatrix}$$

Thus we can take $e_1 = e_2 = 1$. Also $\beta_1(r) = r\alpha$, $\beta_2(n - r) = \beta$.

$$\therefore \ p(N - n, n) = \kappa/\{1\alpha \cdot 2\alpha \dots (N - n)\alpha \cdot \underbrace{\beta \cdot \beta \dots \beta}_{n \text{ terms}}\}$$

$$= \kappa/\{\beta^n \alpha^{N-n}(N - n)!\}$$

Thus

$$\kappa = 1 \bigg/ \left\{ \sum_{n=0}^{N} 1/\{(N - n)!\beta^n \alpha^{N-n}\} \right\}$$

$$\therefore \ p(N - n, n) = \frac{1}{\beta^n \alpha^{N-n}(N - n)!} \bigg/ \left\{ \frac{1}{N!\alpha^N} + \frac{1}{(N-1)!\beta\alpha^{N-1}} \right.$$

$$\left. + \frac{1}{(N-2)!\beta^2\alpha^{N-2}} + \cdots + \frac{1}{\beta^N} \right\}$$

If we multiply the numerator and denominator by $N!\alpha^N$ we obtain

$$p(N-n,n) = \frac{N(N-1)(N-2)\ldots(N-n+1)\left(\dfrac{\alpha}{\beta}\right)^n}{1+N\dfrac{\alpha}{\beta}+N(N-1)\left(\dfrac{\alpha}{\beta}\right)^2+\cdots+N!\left(\dfrac{\alpha}{\beta}\right)^N}$$

which is in accord with the result for p_n in (3.16) and (3.17).

3

$$\mathbf{P} = \begin{bmatrix} 0 & 1 & 0 \\ 0 & 0 & 1 \\ 1 & 0 & 0 \end{bmatrix}$$

so that $\pi_1 = \pi_2 = \pi_3 = 1/3$ and we may take $e_1 = e_2 = e_3 = 1$.

$\beta_1(1) = \beta_1$, $\beta_1(2) = 2\beta_1$, $\beta_1(3) = \beta_1(4) = \beta_1(5) = 2\beta_1$ (two working components), $\beta_2(n_2) = \beta_2$; $\beta_3(n_3) = \beta_3$.

$$\therefore\ p(n_1,n_2,n_3) = \kappa\,\frac{1}{\beta_1(1)\beta_1(2)\ldots\beta_1(n_1)}\left(\frac{1}{\beta_2}\right)^{n_2}\left(\frac{1}{\beta_3}\right)^{n_3}$$

where $n_1 + n_2 + n_3 = 5$.
There are $\binom{7}{2} = 21$ different states; the probabilities are:

$$p(5,0,0) = \kappa\left(\frac{1}{2}\right)^4\left(\frac{1}{\beta_1}\right)^5$$

$$p(4,1,0) = \kappa\left(\frac{1}{2}\right)^3\left(\frac{1}{\beta_1}\right)^4\left(\frac{1}{\beta_2}\right)$$

$$p(4,0,1) = \kappa\left(\frac{1}{2}\right)^3\left(\frac{1}{\beta_1}\right)^4\left(\frac{1}{\beta_3}\right)$$

$$p(3,2,0) = \kappa\left(\frac{1}{2}\right)^2\left(\frac{1}{\beta_1}\right)^3\left(\frac{1}{\beta_2}\right)^2$$

$$p(3,0,2) = \kappa\left(\frac{1}{2}\right)^2\left(\frac{1}{\beta_1}\right)^3\left(\frac{1}{\beta_3}\right)^2$$

$$p(3,1,1) = \kappa\left(\frac{1}{2}\right)^2\left(\frac{1}{\beta_1}\right)^3\left(\frac{1}{\beta_2}\right)\left(\frac{1}{\beta_3}\right)$$

$$p(2,2,1) = \kappa\,\frac{1}{2}\left(\frac{1}{\beta_1}\right)^2\left(\frac{1}{\beta_2}\right)^2\frac{1}{\beta_3}$$

$$p(2,1,2) = \kappa\,\frac{1}{2}\left(\frac{1}{\beta_1}\right)^2\left(\frac{1}{\beta_2}\right)\left(\frac{1}{\beta_3}\right)^2$$

$$p(2,3,0) = \kappa \, \frac{1}{2} \left(\frac{1}{\beta_1}\right)^2 \left(\frac{1}{\beta_2}\right)^3$$

$$p(2,0,3) = \kappa \, \frac{1}{2} \left(\frac{1}{\beta_1}\right)^2 \left(\frac{1}{\beta_3}\right)^3$$

$$p(1,0,4) = \kappa \left(\frac{1}{\beta_1}\right) \left(\frac{1}{\beta_3}\right)^4$$

$$p(1,4,0) = \kappa \left(\frac{1}{\beta_1}\right) \left(\frac{1}{\beta_2}\right)^4$$

$$p(1,3,1) = \kappa \left(\frac{1}{\beta_1}\right) \left(\frac{1}{\beta_2}\right)^3 \left(\frac{1}{\beta_3}\right)$$

$$p(1,1,3) = \kappa \left(\frac{1}{\beta_1}\right) \left(\frac{1}{\beta_2}\right) \left(\frac{1}{\beta_3}\right)^3$$

$$p(1,2,2) = \kappa \left(\frac{1}{\beta_1}\right) \left(\frac{1}{\beta_2}\right)^2 \left(\frac{1}{\beta_3}\right)^2$$

$$p(0,5,0) = \kappa \left(\frac{1}{\beta_2}\right)^5$$

$$p(0,0,5) = \kappa \left(\frac{1}{\beta_3}\right)^5$$

$$p(0,4,1) = \kappa \left(\frac{1}{\beta_2}\right)^4 \left(\frac{1}{\beta_3}\right)$$

$$p(0,1,4) = \kappa \left(\frac{1}{\beta_2}\right) \left(\frac{1}{\beta_3}\right)^4$$

$$p(0,3,2) = \kappa \left(\frac{1}{\beta_2}\right)^3 \left(\frac{1}{\beta_3}\right)^2$$

$$p(0,2,3) = \kappa \left(\frac{1}{\beta_2}\right)^2 \left(\frac{1}{\beta_3}\right)^3$$

κ has to be chosen so that the sum of these probabilities is 1. This in principle is not difficult; just by straightforward addition, for example. The resulting expression is of course cumbersome. The machine cannot operate if $n_1 < 2$. Thus we shall need to find $p(0, 0, 5) + p(0, 1, 4) + p(0, 2, 3) + p(0, 3, 2) + p(0, 4, 1) + p(0, 5, 0) + p(1, 0, 4) + p(1, 4, 0) + p(1, 2, 2) + p(1, 3, 1) + p(1, 1, 3)$.

10
Simulation models

10.1 Introduction

The previous chapters have discussed a number of mathematical models for a number of queueing systems. We have been able to represent the behaviour of these systems by sets of equations which describe the relationships between the variables which describe the systems. We have been able to solve these equations so as to obtain values for these variables in terms of the parameters which describe the systems, the average arrival rate, the number of servers, etc. This in turn has enabled us to calculate many of the important characteristics of the system such as the average number of customers in the system, the proportion of time the server is busy, the mean queueing time of a customer, etc. Furthermore, the results have allowed us to predict what would happen to the system if we changed the parameters. We can calculate the impact of increasing the service rate or of scheduling the arrivals without having to run the system on an experimental basis to see what would happen; a trial and error procedure which could be very costly and inconvenient. The mathematical models have been very powerful.

Of course what has been presented here is but the 'tip of an enormous iceberg'. There is a vast bibliography on queues which has accumulated over the last four or five decades. A glance at journals such as *The Journal of Applied Probability*, *Operations Research*, or *The European Journal of Operational Research*, to name but a few, will reveal many papers on queues. Many researchers have considered many queueing situations, mostly motivated by real world problems, and have constructed mathematical models to describe them. In each case the essence of the methodology has been to capture the behaviour of the system in the form of mathematical relationships. It must be said that in many cases these workers have shown great ingenuity in doing this and in consequence the logic and mathematics of some of the models make great demands on those wishing to understand the work.

Faced with this mountain of interesting literature, it would be easy to believe that we can construct or find a suitable mathematical model for every situation. Unfortunately this is not true. Some systems defy our efforts to make them tractable. In such cases a way forward can be to simulate the system. This still involves developing a model for the system which is such that the model will reproduce, in tangible form, the behaviour of the system. Our simulation model will involve the parameters such as the arrival rate and the number of servers, etc. and will use the values of these parameters to generate realisations of the patterns of behaviour of the system. By changing the parameters we can see how our simulation model responds and can hope to extrapolate this to predict how the actual system will respond.

The details as to how the random variables representing inter-arrival times or service times can be generated are not discussed here. They can be, and have been, for our illustrative purposes, generated on a computer. Indeed for any serious simulation exercise it is necessary to use a computer. As we shall see, a 'hand simulation' is too tedious to carry out for the time needed to get any useful results. In connection with computer simulation it might be mentioned that there are a number of specialist simulation languages (e.g. SIMULA, GPSS, etc.) which have been designed to make the writing of code for simulation programs easy and straightforward. Again, this is too specialised an area for us to investigate.

To illustrate the general procedure we shall consider some simulation models as examples. In some cases there will be a corresponding mathematical model, in which case we shall be able to see just how well the simulation model performs. In general it will be seen that the theoretical model (when it exists) is much more powerful and much more efficient than a simulation model. In that sense, simulation is the approach of last resort.

The broad philosophy of our approach to the simulation of a system is that we shall try to simulate the behaviour of the system over a (long) period of time. During this time certain events happen at certain times which generate changes in the system. In between these events the operation of the system is well defined. In queueing systems the events in question are frequently the arrival of a new customer and the departure of a customer whose service terminates. Our simulation model must keep a record of the important characteristics of the system (how many customers are present, is the server busy? etc.) When the system undergoes a change all the records must be updated. We must also record the time on a clock and the times at which the system changes. The records are then changed, the clock advanced to the new time and all the 'housekeeping' kept in order until the next change occurs. This procedure, shown in Fig. 10.1, goes on for the duration of the simulation.

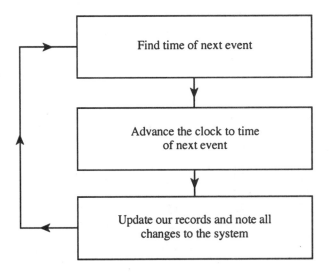

Fig. 10.1

10.2   Hand simulation of an M/M/1 system

We recall that we have already carried out a simulation of an M/M/1 system in Table 2.1 in connection with patients who arrive at the doctor's surgery at random at an average rate of 6/hour (1/10 of a patient per minute), and whose service times have a negative exponential distribution with mean 5 minutes. The inter-arrival times and service times as generated by a computer for a possible realisation of a 2-hour period are given in the Introduction and Chapter 2. Thus, if we recall the work from Chapter 2, the arrival times of the 16 patients are: 11, 12, 14, 19, 21, 22, 30, 39, 45, 49, 67, 71, 78, 87, 115 and 116 minutes. The service times are (in minutes): 2, 3, 10, 3, 2, 1, 5, 4, 11, 5, 3, 10, 7, 15, 5, 1.

Then, bearing in mind that there is a first-in-first-out queue discipline, and that patients are served (seen by the doctor) as soon as possible after they arrive but not before the service completion of the previous patient, we can draw up a table (Table 10.1) showing for each patient their arrival and departure times. We can add to this table and show, for example, the time each patient has to wait before being seen by the doctor, as well as the total time spent in the system for each patient.

$$\text{Time into service} = \text{Max} \begin{cases} \text{Arrival time} \\ \text{Departure time of previous patient} \end{cases}$$

Queueing time = Q_i = Time into service – A_i
Time in system = $T_i = S_i - A_i = Q_i$ + service time.

Our simulation has traced 16 patients through the system. The total patient minutes spent in the system in our realisation is 131. It is the area under the graph of $N(t)$ against t as given in Chapter 2. Thus the mean time a patient spends in the system is

Table 10.1

Patient number (i)	Arrival time A_i	Time into service	Service time	Departure time S_i	Queueing time Q_i	Time in system T_i
1	11	11	2	13	0	2
2	12	13	3	16	1	4
3	14	16	10	26	2	12
4	19	26	3	29	7	10
5	21	29	2	31	8	10
6	22	31	1	32	9	10
7	30	32	5	37	2	7
8	39	39	4	43	0	4
9	45	45	11	56	0	11
10	49	56	5	61	7	12
11	67	67	3	70	0	3
12	71	71	10	81	0	10
13	78	81	7	88	3	10
14	87	88	15	103	1	16
15	115	115	5	120	0	5
16	116	120	1	121	4	5
Total					44	131

$131/16 \approx 8.19$ minutes. The theoretical value from (2.14) is $1/(1/5 - 1/10) = 10$ minutes. The mean number of patients in the system is $131/121 \approx 1.1$. The divisor is 121 since our simulation extends over 121 minutes. Again the theoretical value is, from (2.10) with

$$\rho \left(= \frac{\alpha}{\beta} = \frac{\dfrac{1}{10}}{\dfrac{1}{5}} \right) = \frac{1}{2}$$

given by $1/2/(1 - 1/2) = 1$.

Thus our simulation has given reasonable results but they are not spectacularly accurate, basically because we have not simulated the system for long enough. In our simulation the mean inter-arrival time for the 16 patients is $116/16 = 7.25$ minutes and the mean service time $= 87/16 = 5.44$ minutes and these differ from the theoretical expected values of 10 and 5 minutes, respectively. We might therefore question just how typical our realisation is. This is always a problem with any simulation exercise. The random variables generated are just a random sample from their respective distributions and as such the usual sampling errors are bound to occur. There are procedures which can and should be applied in any serious simulation exercise to help overcome these difficulties, but they are not discussed here.

10.3 Computer simulation of a G/G/1 queue and a G/G/2 queue

In order to carry out a sensible simulation of a single-server queue it is necessary to use a computer. The operation of the system follows a logical pattern which repeats itself over and over again, as successive customers arrive and depart, and this logic can be coded as a computer program. We have indicated in Fig. 10.2 the form of the model. The abbreviations are in mnemonic form. The next event which changes the system is either an arrival or a service completion. The computer will have to store and update the next event time (NET), be that the next arrival time (NAT) or the next service completion time (NSCT). The computer generates inter-arrival times (IAT) and duration of service (i.e. service) times (DS), and so if appropriate distributions are used to generate these values the logic of the diagram applies to a G/G/1 system. The various quantities which the computer records and updates as the realisation unfolds are listed in Table 10.2. Details of the code are not given, but the program has been well tested. The program will simulate CS service completions. Output concerning the main quantities of interest is printed after every CO service completions. CS and CO are at our choice and can be chosen so as to show the convergence of the system to the steady-state results.

We reproduce the output from a run of 3000 service completions for an M/M/1 queue. The subroutines to generate the inter-arrival times (IAT) and service times (DS) were written to generate negative exponential variables with mean $1/\lambda$ and $1/\mu$, respectively.

The first number input seeds the computer's random number generator. Just one simulation of a single-server queue is carried out with $\lambda = 0.8$, $\mu = 1$ so that the traffic intensity $\rho = \lambda/\mu = 0.8$. A total of 3000 service completions are simulated with output being given after each set of 300 service completions. There were no customers present

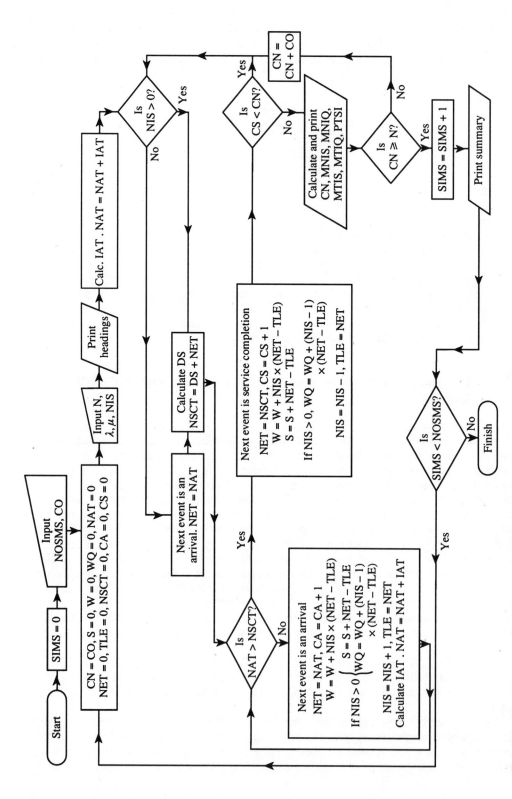

Fig. 10.2 Flow chart for single-server queue simulation – first-come-first-served discipline

Table 10.2

Key	
CA	Counts arrivals
CN	A counter to control output
CO	Output is performed every CO services
CS	Counts services
DS	Duration of service (Service time)
IAT	Inter-arrival time
LAM(BDA)	Arrival rate (λ)
MNIQ	Mean number in queue
MNIS	Mean number in system
MTIQ	Mean time in queue
MTIS	Mean time in system
MU	Service rate (μ)
N	Number of services per simulation
NAT	Next arrival time
NET	Next event time
NIS	Number in the system
NOSMS	Number of simulations to be done
NSCT	Next service completion time
NSVS	Number of services
PTSI	Proportion of time server idle
RHO	Traffic intensity ($=\lambda/\mu$)
S	Serving time
SIMS	A counter for number of simulations
TLE	Time of last event
W	Waiting time in system
WQ	Waiting time in queue

in the system at the outset. Figure 10.3 shows that during the simulation there were 3007 arrivals. There will have been seven left in the system after the service completion of the 3000th customer.

From (2.10) the theoretical value of the mean number in the system ($\rho/(1-\rho)$) is 4 when $\rho = 0.8$, as compared to our last estimate of 3.5. For the mean time in the system the theoretical value from (2.14) ($1/(\beta - \alpha)$) is 5 minutes, as compared to 4.36 minutes by simulation. For the proportion of time the server is idle the theoretical value from (2.8) is $(1 - \rho) = 0.2$, as compared to 0.203 from the simulation. Thus our estimates from simulation are not particularly accurate and fluctuate quite markedly as the simulation progresses. Yet an enormous amount of work has been done (by the computer) to get these results. The truth is that it is very time consuming and difficult to obtain accurate results from a simulation. We repeat, if a mathematical model exists and can be used it will generally be more accurate and more convenient than a vast simulation exercise.

By changing the routine which generates the service time (DS) in the program, to one in which DS is always assigned the constant value $1/\mu$, we can simulate the M/D/1 system. We shall use exactly the same arrival rate and service rate as was used in the M/M/1 system, but this time all variation in the service time has been eliminated. We

```
RANDOM GENERATOR ? 1749
ENTER NOSMS ? 1
ENTER COUNT CONTROL ? 300
NO. OF SERVICES ? 3000
ARRIVAL RATE ? 0.8
SERVICE RATE ? 1
  TRAFFIC INTENSITY = .8
INITIAL NO. IN SYSTEM ? 0
SINGLE-SERVER QUEUE SIMULATION
NSVS    MNIS    MNIQ    MTIS    MTIQ    PTSI
 300    3.853   3.005   4.334   3.380   0.152
 600    5.744   4.879   6.437   5.468   0.135
 900    5.533   4.685   6.402   5.420   0.152
1200    4.637   3.804   5.592   4.588   0.167
1500    4.186   3.367   5.041   4.054   0.181
1800    4.021   3.205   4.849   3.866   0.184
2100    3.739   2.939   4.589   3.607   0.200
2400    3.593   2.799   4.466   3.479   0.206
2700    3.558   2.760   4.406   3.418   0.202
3000    3.503   2.706   4.362   3.370   0.203
NUMBER OF ARRIVALS = 3007
DURATION OF SIMULATION = 3744.553
SERVING TIME = 2982.634
```

Fig. 10.3

start with an empty system as before and show the important characteristics of the system over the duration of our simulation, again for 3000 service completions (Fig. 10.4).

Reducing the variation in service time does not appear to have had much effect on the proportion of time the server is idle (now 0.207 as compared with 0.203 before). This, of course, is in line with the theory $[1 - \rho = 1 - \alpha/\beta(\lambda/\mu$ in the program)] from (2.6) and (4.4) with $\rho = 0.8$ in these cases. However, the mean number in the system has been reduced from 3.5 to 2.2 (to 1 decimal place). The theoretical values are from (2.10)

$$\left(\frac{\rho}{1-\rho}\right) = 4.0 \qquad \text{for the M/M/1 case}$$

and from the discussion just following (4.5)

$$\frac{\rho}{1-\rho}\left(1 - \frac{1}{2}\rho\right) = 2.4 \qquad \text{for the M/D/1 case}$$

The simulation model is indicating the way the system will change in the circumstances. However, it is only giving a vague and not spectacularly accurate indication. It is much more precise and easy to use the formulae from the mathematical models, and for them we can change the value of ρ in a trice and still come up with the correct predicted values with minimal effort. All this reinforces the power of mathematical models (when they exist) and their superiority over simulation models.

In case the reader has any doubts concerning what happens to a single-server queue if the arrival rate is actually equal to the service rate ($\rho = 1$), we show the outcome of a simulation of an M/M/1 system for 10000 service completions. It is readily seen that the number in the system and the waiting times are increasing, whilst at the same time the server is kept fully occupied. There is no steady state in this case (Fig. 10.5).

The logic of the flow chart for the simulation of the G/G/1 system can easily be

```
RANDOM GENERATOR ? 4937
ENTER NOSMS ? 1
ENTER COUNT CONTROL ? 300
NO. OF SERVICES ? 3000
ARRIVAL RATE ? 0.8
SERVICE RATE ? 1
  TRAFFIC INTENSITY = .8
INITIAL NO. IN SYSTEM ? 0
SINGLE-SERVER QUEUE SIMULATION
NSVS    MNIS    MNIQ    MTIS    MTIQ    PTSI
 300    2.599   1.768   3.055   2.078   0.169
 600    2.410   1.614   3.028   2.028   0.204
 900    2.447   1.638   3.024   2.025   0.192
1200    2.225   1.429   2.791   1.793   0.204
1500 .  2.131   1.343   2.699   1.701   0.212
1800    2.150   1.357   2.706   1.707   0.206
2100    2.107   1.311   2.645   1.646   0.204
2400    2.082   1.287   2.616   1.617   0.205
2700    2.032   1.245   2.579   1.581   0.213
3000    2.184   1.392   2.753   1.754   0.207
NUMBER OF ARRIVALS = 3002
DURATION OF SIMULATION = 3784.013
SERVING TIME = 3000.005
```

Fig. 10.4

```
RANDOM GENERATOR ? 6727
ENTER NOSMS ? 1
ENTER COUNT CONTROL ? 1000
NO. OF SERVICES ? 10000
ARRIVAL RATE ? 1
SERVICE RATE ? 1
  TRAFFIC INTENSITY = 1
INITIAL NO. IN SYSTEM ? 0
SINGLE-SERVER QUEUE SIMULATION
NSVS    MNIS    MNIQ    MTIS    MTIQ    PTSI
 1000   40.584  39.591  40.309  39.323  0.007
 2000   47.682  46.686  47.552  46.559  0.003
 3000   44.910  43.912  44.871  43.874  0.002
 4000   46.605  45.606  46.761  45.759  0.002
 5000   51.053  50.055  51.154  50.154  0.001
 6000   62.646  61.647  62.601  61.603  0.001
 7000   74.857  73.858  74.538  73.543  0.001
 8000   83.179  82.180  82.887  81.891  0.001
 9000   87.122  86.123  87.164  86.164  0.001
10000   90.184  89.185  90.239  89.239  0.001
NUMBER OF ARRIVALS = 10130
DURATION OF SIMULATION = 10136.18
SERVING TIME = 10129.19
```

Fig. 10.5

amended to deal with the $G/G/m$ system; a queue with general inter-arrival times, general service time and m identical servers with a first-in-first-out queue discipline.

The record-keeping needs to be a bit more elaborate. For each server we need to know whether that server is busy or not and when that server will next become free. The earliest of these times will generate the next service completion time for such a system.

The output shown in Fig. 10.6 is from the simulation of 4000 service completions in an M/M/2 system with (in the notation of Chapter 3) $\alpha = 1.6$ (λ of the program) and

```
RANDOM GENERATOR ? 4197
NUMBER OF SERVERS ? 2
NO. OF SIMULATIONS ? 1
COUNT CONTROL ? 400
NUMBER OF SERVICES ? 4000
ARRIVAL RATE ? 1.6
SERVICE RATE ? 1
INITIAL NO. IN SYSTEM ? 0
MULTI-SERVER QUEUE SIMULATION
LAMBDA = 1.6 MU = 1 RHO = 1.6
NUMBER OF SERVERS = 2
NSVS    MNIS    MNIQ    MTIS    MTIQ    PTSB
 400    2.820   1.308   1.710   0.793   0.756
 800    3.372   1.818   2.053   1.107   0.777
1200    4.892   3.232   2.970   1.962   0.830
1600    4.225   2.636   2.632   1.642   0.794
2000    4.216   2.606   2.598   1.606   0.805
2400    4.055   2.447   2.511   1.516   0.804
2800    3.999   2.387   2.465   1.471   0.806
3200    3.847   2.253   2.396   1.403   0.797
3600    3.774   2.187   2.371   1.374   0.793
4000    4.044   2.444   2.535   1.532   0.800
NO. OF ARRIVALS = 4008
DURATION OF SIMULATION = 2512.302
```

Fig. 10.6

$\beta = 1$ (μ of the program) so that $\rho = 1.6$, which we note is less than 2 which it has to be for a steady-state solution to exist (3.6).

By (3.7) the theoretical value for L is given by $4\rho/(4 - \rho^2) = 4.44$ (simulated value 4.044). The proportion of time a server is busy is given by $1 - (p_0 + \frac{1}{2}p_1)$. This last result assumes that a customer arriving when the system is empty chooses a server at random. Now from (3.6) $p_0 = (2 - \rho)/(2 + \rho) = 1/9$ when $\rho = 1.6$ and so $p_1 = \rho p_0 = 1.6/9$ in this case. Thus $p_0 + \frac{1}{2}p_1 = 1.8/9 = 0.2$. Thus our simulation gives a very accurate estimate for this quantity. The reader is encouraged to compare other estimates from the simulations with the corresponding theoretical results from the M/M/1 and M/M/2 models.

10.4   The simulation of a machine-minding problem

The machine interference model, which was discussed in Chapter 3 as a special case of a birth–death model, did not take account of the time taken by the operator to walk to a stopped machine. The problem now considered includes this feature. Suppose the operator looks after N identical machines which break down at random in running time at rate B for each machine. Suppose the machines are laid out on the factory floor in two lines with equal spacing between the lines and between the machines in a line so that it takes a time W for the operator to walk from one machine to the next and inspect the next machine for faults (Fig. 10.7). If a fault is found, the machine will be stopped and the operator then spends an additional time R repairing the machine. The operator's mode of working is to patrol the machines uni-directionally and repair stopped machines when they are encountered. This extends the machine interference problem of Chapter 3. A machine may have to wait for attention not simply because the operator is repairing another machine when it stops, but also because the operator is elsewhere on the patrol when it stops, and takes a finite time to walk to the stopped machines. It might be felt that by taking account of the walking time, assuming random breakdowns and constant repair times (for a routine repair), that this is a more realistic model for this problem.

A theoretical solution to this problem has been given but we shall study the system by simulation. With the assumption of random breakdowns, the run time of a machine

$N = 10$

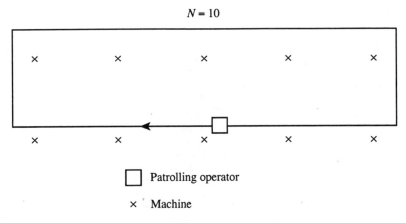

☐ Patrolling operator

× Machine

Fig. 10.7

from repair until the next breakdown has a negative exponential distribution with mean $1/B$ (0.3). The simulation model estimates the efficiency of the machines (the ratio of the actual running time achieved to that which would have been obtained with no stops). This and other quantities of interest are output at regular intervals of time. The system is simulated for a certain period of time, called a shift time, and this can be determined by the user of the program.

The logic underlying the model (along with a key describing the quantities involved in Table 10.3) is shown in Fig. 10.8.

With $N = 8$, $R = 1$ minute, $B = 1/20$, and $W = 1/4$ minute, and with the simulation covering a period of 3600 minutes, the output shown in Fig. 10.9 was obtained. Intermediate output has been given every 200 minutes and shows the convergence to the steady-state value of the efficiency.

Table 10.3

Key	
B	Breakdown rate in running time per machine
BR	Product of breakdown rate and repair time
BT(I)	Breakdown time for machine I (time it next stops)
CT	Intermediate output is printed every CT time units (approx.)
EFF	Machine efficiency as a percentage
IDT	Idle time of all the machines
N	Number of machines
NBW	Product of number of machines, breakdown rate and walking time
NINS	Number of inspections of machines
NRP	Number of repairs done
OPT	Operator's time (actual time)
R	Repair time
Run-time	The time for a machine to run from a repair to next breakdown
SH	Shift time (duration of simulation)
T	Total idle time, including the waiting time of those machines which have not yet been repaired
W	Walking time from machine I to machine I + 1

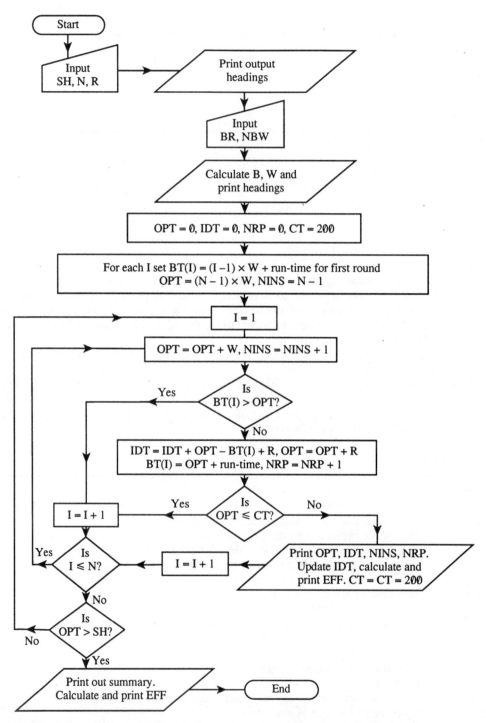

Fig. 10.8 Flow chart for simulation of *N* machines uni-directionally patrolled by one operator when walking time and repair time are constant. Breakdowns occur at random so that run-time has a negative exponential distribution

```
RANDOM GENERATOR ? 3135
SHIFT-TIME? 3600
NO. OF MACHINES? 8
REPAIR TIME? 1
MACHINE EFFICIENCY,UNI-DIRECTIONAL PATROLLING
CONSTANT  WALKING  TIME  (W),CONSTANT  REPAIR  TIME
(R)
BR? 0.05
NBW? 0.1
N = 8  BR = .05  NBW = .1
OPTIME    IDTIME    NINS    NRP     EFF%
 201.00   153.31    548      64    90.369
 401.00   355.83   1060     136    88.908
 600.50   515.22   1602     200    89.275
 813.25   732.70   2153     275    88.698
1001.75   911.09   2643     341    88.619
1200.75  1145.37   3127     419    88.077
1401.75  1272.26   3715     473    88.655
1600.50  1446.43   4242     540    88.703
1801.00  1627.47   4776     607    88.685
2000.25  1813.58   5293     677    88.667
2204.00  2005.73   5812     751    88.625
2401.00  2162.99   6344     815    88.739
2600.75  2299.78   6931     868    88.939
2801.75  2494.67   7443     941    88.857
3001.25  2718.27   7933    1018    88.675
3200.25  2893.55   8449    1088    88.698
3400.75  3114.34   8959    1161    88.553
SUMMARY
SH = 3600  N = 8
B = .05  R = 1  W = .25
BR = .05  NBW = .1
EFFICIENCY = 88.45516 %
```

Fig. 10.9

The mathematical model for this system proves that the efficiency depends on the parameters N, B, R and W only through the products BR and NBW. The theoretical value of the efficiency for the parameter values used in our simulation is 88.1, so our simulation is quite good (and consistent) in this particular case.

10.5 Some general comments

The simulations described in this section can do no more than give a flavour of the general method. Numerous refinements can and should be incorporated into simulation models for queueing systems models and we have not discussed such features. They could form a small course in their own right. The general conclusion is that when it is not possible to use a theoretical mathematical model then simulation is always a fall-back position. It should, however, only be used as the last resort, since in general it will involve a lot of work for a not particularly accurate result.

10.6 Examples 20

1 Use the results from the simulation of the doctor's surgery to estimate the mean time spent waiting in the surgery by a patient. Compare this with the theoretical result for the $M/M/1$ system.

2 Use the graphical representation of $N(t)$ against t for the doctor's surgery (Table 2.1) to estimate the proportion of time the doctor is busy and the mean length of the doctor's busy periods.

3 Suppose the 16 patients had been scheduled to arrive at 8-minute intervals so that the arrival times are at 0, 8, 16, 24, 32, ..., 120. Repeat the simulation using the same service times (times spent with the doctor) and now estimate:

  (i) the mean waiting time;
  (ii) time in the system for a patient;
  (iii) the proportion of time the doctor is not seeing patients; and
  (iv) the probability that an arriving patient does not have to wait.

Compare these outcomes with the theoretical results.

4 A production line turns out about 100 items per day, but deviations from this can occur due to a number of causes and the production is more accurately described by the probability distribution:

No. of items	95	96	97	98	99	100	101	102	103	104	105
Probability	0.03	0.05	0.07	0.10	0.15	0.20	0.15	0.10	0.07	0.05	0.03

Production from the line is sent to a despatch bay and at the end of each day the items at the bay are transported by conveyance to another part of the factory. This conveyance can only carry 101 items and numbers in excess of this are left behind until the next day.

Simulate the system over a 40-day period, say, to investigate the average number of items left waiting at the end of each day and the average number of empty spaces on the conveyance.

5 Simulate the machine interference model of Chapter 3 with four machines and one operator with, say, $\alpha = 1/100$ and $\beta = 1/10$. Estimate the average number of machines running.

Repeat the exercise above where a team of two operators looks after eight machines. Is the efficiency higher in this case? Why should that be? Consider how the logic of the operations could be represented on a flow chart ready for conversion to a computer program.

6 Try to consider in the same way the problem of simulating the systems in, for example: Examples 17, question 2; Examples 18, question 2; and Examples 19, question 3.

10.7 Solutions to Examples 20

1 From Table 10.1, the 16 patients wait before seeing the doctor for a total of 44 patient minutes. Thus the mean waiting time $= 44/16 = 2.75$ minutes.

The theoretical value (2.13) is given by

$$W_q = \frac{\alpha}{\beta(\beta - \alpha)}$$

(with $\alpha = 1/10$ and $\beta = 1/5$ in this case) and so is 5 minutes.

2 From the graph of $N(t)$ against t in Fig. 2.5 for this realisation, the doctor is busy for a total of 87 minutes out of 121 minutes. Therefore, proportion of time doctor is busy = $87/121 \approx 0.72$, as compared with the theoretical value of 0.5 ($=\rho$) from (2.9). This 87 minutes comprises six busy periods whose average length is thus 14.5 minutes. From (2.16) the theoretical value for the mean length of a busy period is 10 minutes in this case.

We should not be alarmed at the apparent inaccuracies just noted. The theoretical results hold for the steady-state solution and our simulation has not been carried out over a sufficiently long time. In any case one could argue that the steady-state solution cannot be applied to the doctor's surgery problem. This is always an event that lasts for approximately 2 hours and each occurrence is discrete and not a continuation of the previous surgery. Thus repeated simulation of 2-hour periods starting with an empty surgery followed by averaging of the results might be the way to get better estimates for the quantities concerned.

3 Since the doctor (perhaps) knows that he sees patients for about 5 minutes on average, to schedule their arrival times at 8-minute intervals is not unreasonable (perhaps a practical implementation would be two every quarter of an hour).

With the assumption of arrivals at 0, 8, 16, ..., 120 we can draw up a table (Table 10.4) for a realisation of the surgery with the same service times as those used earlier.

$$\text{(Time into service)}_i = \text{Maximum}\{A_i, S_{i-1}\}$$
$$Q_i = \text{(Time into service)}_i - A_i$$
$$T_i = S_i - A_i = Q_i + \text{(Service time)}_i$$

Table 10.4

Patient number (i)	Arrival time A_i	Time into service	Service time	Departure time S_i	Queueing time Q_i	Time in system T_i
1	0	0	2	2	0	2
2	8	8	3	11	0	3
3	16	16	10	26	0	10
4	24	26	3	29	2	5
5	32	32	2	34	0	2
6	40	40	1	41	0	1
7	48	48	5	53	0	5
8	56	56	4	60	0	4
9	64	64	11	75	0	11
10	72	75	5	80	3	8
11	80	80	3	83	0	3
12	88	88	10	98	0	10
13	96	98	7	105	2	9
14	104	105	15	120	1	16
15	112	120	5	125	8	13
16	120	125	1	126	5	6
Total			87		21	108

From the simulation our estimates for the quantities stated are:

  (i) Mean time waiting to see the doctor $= 21/16 \approx 1.31$ minutes
 (ii) Mean time in the system $= 108/16 = 6.75$ minutes
(iii) Proportion of time the doctor is not seeing patients $= (126 - 87)/126 \approx 0.31$
(iv) Probability that a patient does not have to wait $= 10/16 = 5/8 = 0.625$ since 10
       out of 16 patients go straight into the doctor in our realisation.

In view of the remarks made following the solution to question 2 we should be hesitant in comparing these to the theoretical results for the D/M/1 system given in Chapter 5. In the notation of that section, $\beta = 1/5$ and the inter-arrival time is constant at 8 minutes. Thus from (5.7) and the solution of question 2 of Examples 12, the equation that determines η is $z = \exp\{-\frac{8}{5}(1 - z)\}$, hence $\eta \approx 0.358$. Thus from the solution to question 1 of Examples 12, the mean time spent in the queue

$$\left(\frac{\eta}{\beta(1 - \eta)}\right)$$

is 2.79 minutes, approximately, whilst the mean time spent in the system

$$\left(\frac{1}{\beta(1 - \eta)}\right)$$

is 7.79 minutes

    For (iii) there is no theoretical result that we have obtained and for (iv) our calculated value of u_0, the probability that an arriving customer finds an empty system $(1 - \eta)$, is 0.642.

    We can, as in Chapter 2, represent the behaviour of the system over the 2-hour period in graphical form by plotting $N(t)$, the number of patients at time t, against t. This we have done. There is a small problem at times 80 and 120 when, due to our working in discrete minutes, we have a simultaneous arrival and departure (not allowed in theory with continuous time).

    Figure 10.10 shows that the effect of scheduling the arrivals is to reduce the congestion and decrease the mean length of a busy period (we have no theoretical result for this last quantity for a D/M/1 system). In this case the total time for which

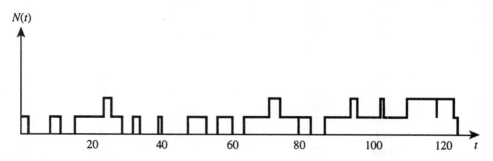

Fig. 10.10

the doctor is busy, namely 87 minutes, comprises 10 busy periods (we have assumed a break at 80) with mean length 8.7 minutes. The scheduling also increases the proportion of patients who find an empty system and so do not have to queue. For an M/M/1 system with $\rho = \alpha/\beta = 5/8$ in this case, this is by (2.8) $p_0 = 1 - \rho = 3/8 = 0.375$.

In theory for a D/M/1 system with $\rho = 5/8$ it is $u_0 = 1 - \eta$ by (5.8) and so is 0.642 since we have shown η to be 0.358. This is a significant increase and shows the value of the scheduling exercise. Our simulation has yielded 0.625 as an estimate in this case. However, as we have said before, since the doctor's surgery only operates for about 2 hours at a time it is inappropriate to use the steady-state solution. The transient solution should be used and we certainly do not have a mathematical model and solution for that.

If we look back over the outcomes for our solutions to questions 1, 2 and 3 of these examples, we see the virtue of reducing variation in the arrival pattern and the service provision. The D/M/1 system causes less congestion than the M/M/1 system. Had the doctor been able to allocate exactly 5 minutes to each of his patients who arrived on schedule at 8-minute intervals, then for that D/D/1 system there would be no congestion at all [see question 1 of Examples 5].

4 This 'simple' problem does not come within the scope of our mathematical models. The flow chart in Fig. 10.11 outlines the form of a simulation of the system.

Based on this logic we show the simulation of 40 days of activity (Fig. 10.12). The realisation of the numbers made is compared with the expected numbers made, were these to conform exactly to the distribution.

5 In outline we need to find for each machine the next time it will break down; BT(I) for machine I. This in turn will be derived from a realisation of the run time for that machine, RT(I), which is added to the time at which it is set running. The time at which the operator commences the next repair, which is on machine J, is BT(J), where BT(J) = Min$_I$ BT(I), and this determines J. Time is then updated to $T = $ BT(J) + R where R is the (generated) repair time and BT(J) reset to $T + $ RT(J) where RT(J) is a run time.

Records of the number of repairs, time to carry out the repairs, time for which each machine is stopped (and running) are kept. Then the time at which the next repair commences is determined as above, etc.

For the case of one operator, four machines, and $\rho = \alpha/\beta = 0.1$, from Table 3.1 we have that the average number of machines running is 3.53.

When a team of two operators looks after eight machines it can be shown that the theoretical value of the average number of machines running is 7.21. [It is not so difficult. In the notation of Chapter 3 we have birth–death equations with $\alpha_n = \alpha(8 - n)$ for $n = 0, 1, \ldots, 8$, $\beta_1 = \beta$ and $\beta_n = 2\beta$ when $n = 2, \ldots, 8$ and both operators are working.]

We note that $7.21 > 2 \times 3.53$ (= 7.06), so the efficiency is higher in this case. It is better to let the operators work as a team on all eight machines rather than assign four machines to each operator. This latter system could lead to situations in which one operator is idle while the other has two or more stopped machines. It might be quite difficult for a simulation solution to pick out the relatively small improvement that has been obtained.

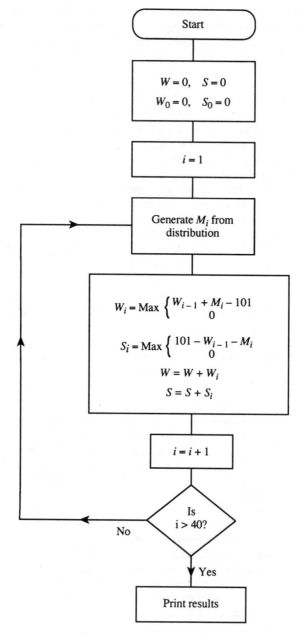

Fig. 10.11 Key: W = total of those left waiting; S = total of spaces left; M_i = number made on day i; W_i = number left waiting on day i; S_i = number of empty spaces on day i

6 The simulation of a network of queues can rapidly become very complicated. We have outlined in flow chart form the operation of a G/G/1 queue (Fig. 10.2). The operation of a network is that of each system appropriately placed. The departure times from one system will of course correspond to the arrival times at the next system.

```
SIMULATION
DAY    ITEMS MADE    SPACES    WAITING
  1       101          0          0
  2       101          0          0
  3        98          3          0
  4       104          0          3
  5       101          0          3
  6       100          0          2
  7        98          1          0
  8        95          6          0
  9       104          0          3
 10        99          0          1
 11        99          1          0
 12        97          4          0
 13       103          0          2
 14       104          0          5
 15        97          0          1
 16       102          0          2
 17        98          1          0
 18       103          0          2
 19       100          0          1
 20        99          1          0
 21       101          0          0
 22        99          2          0
 23       100          1          0
 24       100          1          0
 25       100          1          0
 26        96          5          0
 27        95          6          0
 28       100          1          0
 29       100          1          0
 30       100          1          0
 31        99          2          0
 32        99          2          0
 33       104          0          3
 34        98          0          0
 35        96          5          0
 36       100          1          0
 37        97          4          0
 38       101          0          0
 39       101          0          0
 40        99          2          0
AVE. NO. OF EMPTY SPACES = 1.3
AVE. NO. OF ITEMS WAITING = .7
HIT A KEY TO COMPARE SIMULATION WITH EXPECTED
ITEMS MADE    OBSERVED    EXPECTED
  95             2           1
  96             2           2
  97             3           3
  98             4           4
  99             7           6
 100             9           8
 101             6           6
 102             1           4
 103             2           3
 104             4           2
 105             0           1
```

Fig. 10.12

Appendix. The transient solution for the M/M/1 queue: a simple solution at last

Introduction

The system of differential-difference equations for a single-server queue, with random arrivals at rate α and exponential service at rate β with a first-in-first-out queue discipline, has already been mentioned (equations (8.1)). If $p_n(t)$ denotes the probability that there are n customers in the system at time t then

$$\frac{dp_0(t)}{dt} = -\alpha p_0(t) + \beta p_1(t)$$

$$\frac{dp_n(t)}{dt} = \alpha p_{n-1}(t) - (\alpha + \beta)p_n(t) + \beta p_{n+1}(t) \qquad \text{for } n \geqslant 1 \tag{A1}$$

If the system is empty at the outset, then we have to find a solution of the equations (A1) subject to the initial conditions

$$p_0(0) = 1, \; p_n(0) = 0, \qquad \text{for } n \neq 0 \tag{A2}$$

These equations were known to the Danish mathematician Erlang (Brockmeyer *et al.*, 1948), who pioneered work on queueing theory in connection with applications to the design of automatic telephone exchanges. This would have been in the 1920s. However, the first solution to these equations did not appear until the 1950s, three decades later (Clarke, 1953; Bailey, 1954; Champernowne, 1956). The solution is the formidable expression (8.2). In fact, if one uses properties of the modified Bessel functions, it is possible to tidy up that expression and obtain $p_n(t)$ in the form

$$p_n(t) = \frac{e^{-(\alpha + \beta)t}\left(\dfrac{\alpha}{\beta}\right)^n}{\beta t} \sum_{k=n+1}^{\infty} \left(\frac{\alpha}{\beta}\right)^{-k/2} kI_k(2\sqrt{\alpha\beta t}) \tag{A3}$$

Here $I_n(z)$ is the modified Bessel function which can be written as

$$I_n(z) = \sum_{m=0}^{\infty} \frac{\left(\dfrac{z}{2}\right)^{n+2m}}{m!(m+n)!} \tag{A4}$$

This last result is still very complicated, as were the methods used to derive it. In the words of Kleinrock (1975), one of the key investigators in queueing theory, 'This last expression is most disheartening. What it has to say is that an appropriate model for the simplest interesting queueing system leads to an ugly expression for the time dependent behaviour of its state probabilities'.

In his book on Markovian queues, Sharma (1990) gives an alternative expression for $p_n(t)$. His work is based on investigating the number of arrivals and the number of service completions over the period $(0, t)$ for an M/M/1 queue. The difference will give the number of customers in the system at time t, assuming that we start with an empty system. His result, which clearly shows the steady-state solution (2.6) if $\rho = \alpha/\beta < 1$, is:

$$p_n(t) = \left(1 - \frac{\alpha}{\beta}\right)\left(\frac{\alpha}{\beta}\right)^n + e^{-(\alpha+\beta)t}\left(\frac{\alpha}{\beta}\right)^n \sum_{m=0}^{\infty} \frac{(\alpha t)^m}{m!} \sum_{k=0}^{m+n} (m-k)\frac{(\beta t)^{k-1}}{k!} \tag{A5}$$

The equivalence of (A3) and (A5) has been established by Conolly and Langaris (1993), although the work is not trivial.

A simple form for the transient solution

In the equations (A1) we replace α/β by ρ and transform to a new time scale $\tau = \beta t$ in which the mean service time ($t = 1/\beta$) is the unit of time. Then since

$$\frac{d}{dt} = \frac{d}{d\tau} \cdot \frac{d\tau}{dt} = \beta \frac{d}{d\tau}$$

the equations (A1) or (8.1) become

$$\frac{d}{d\tau} p_0(\tau) = -\rho p_0(\tau) + p_1(\tau)$$

$$\tag{A6}$$

$$\frac{d}{d\tau} p_n(\tau) = \rho p_{n-1}(\tau) - (1+\rho)p_n(\tau) + p_{n+1}(\tau) \qquad \text{for } n \geq 1$$

The equations (A6) only involve one parameter ρ and this makes things a bit simpler. By replacing ρ by α/β and τ by βt in any solution we can easily recover the solution to (A1) in terms of α and β and t.

Guided by the form of (A3) and (A5) we seek a solution to the equations (A6) of the form

$$p_n(\tau) = e^{-(1+\rho)\tau} \rho^n \sum_{m=0}^{\infty} \frac{a(m,n)\tau^m}{m!} \tag{A7}$$

Then

$$\frac{dp_n(\tau)}{d\tau} = -(1+\rho)e^{-(1+\rho)\tau}\rho^n \sum_{m=0}^{\infty} \frac{a(m,n)\tau^m}{m!} + e^{-(1+\rho)\tau}\rho^n \sum_{m=1}^{\infty} \frac{a(m,n)\tau^{m-1}}{(m-1)!}$$

so that if we substitute these expressions for $p_n(\tau)$ and $dp_n(\tau)/d\tau$ into (A6), after a little simplification, by equating the coefficients of $\tau^m/m!$ in the resulting expansions, we obtain:

$$a(m+1, 0) = a(m, 0) + \rho a(m, 1)$$
$$a(m+1, n) = a(m, n-1) + \rho a(m, n+1) \qquad \text{for } n \geq 1, \tag{A8}$$

for each value of $m = 0, 1, 2, \ldots$.

The initial conditions (A2) translate into the conditions

$$a(0, 0) = 1, \ a(0, n) = 0 \qquad \text{for } n \neq 0 \tag{A9}$$

This gives the values of $a(m, n)$ for $m = 0$. We can then use the two equations (A8) to find in turn $a(1, 0)$ and $a(1, n)$, and these in turn to find $a(2, 0)$ and $a(2, n)$, etc. The outcome of these calculations shows that $a(m, n)$ is a polynomial in ρ whose form for the first few values of m and n is as shown in Table A1.

By inspection we see that $a(m, n) = 0$ for $m < n$, and $a(m, n) = 1$ for $m = n$. For $m > n$ (indeed for $m \geq n$), we observe that $a(m, n)$ is a polynomial in ρ of degree $[(m - n)/2]$ where $[x]$ represents the largest positive integer which is less than x. Furthermore, the coefficient of ρ^r in this polynomial is

$$\binom{m}{r} - \binom{m}{r-1}$$

where

$$\binom{m}{r}$$

is the usual binomial coefficient defined when m is an integer and $0 \leq r \leq m$. Thus we can write

$$a(m, n) = \begin{cases} 0 & \text{for } m < n \\ 1 & \text{for } m = n \\ \displaystyle\sum_{r=0}^{[(m-n)/2]} \left\{ \binom{m}{r} - \binom{m}{r-1} \right\} \rho^r & \text{for } m > n \end{cases} \tag{A10}$$

It should be noted in the last form that if $n > m$ then $(m - n)/2$ is negative so that $[(m - n)/2]$ does not exist and there are no terms in the sum. In that sense all the cases are covered by the last form in (A10) since

$$\binom{m}{0} - \binom{m}{-1} = \binom{m}{0} = 1$$

so that the sum gives the single term 1 when $m = n$.

It is clear that (A10) holds for $m, n = 0, 1, 2$. We can then prove it true generally by induction on m for if

$$a(m, n) = \sum_{r=0}^{[(m-n)/2]} \left\{ \binom{m}{r} - \binom{m}{r-1} \right\} \rho^r$$

up to a certain value m, then for the next value of m we can use (A8). In the first equation, on the R.H.S., we have

$$a(m, 0) + \rho a(m, 1) = \sum_{r=0}^{[m/2]} \left\{ \binom{m}{r} - \binom{m}{r-1} \right\} \rho^r + \sum_{r=0}^{[(m-1)/2]} \left\{ \binom{m}{r} - \binom{m}{r-1} \right\} \rho^{r+1}$$

Now, if m is even, $m = 2k$,

$$\left[\frac{m}{2} \right] = k, \qquad \left[\frac{m-1}{2} \right] = k - 1, \qquad \left[\frac{m+1}{2} \right] = k$$

Table A1 $a(m, n)$

n

m	0	1	2	3	4	5	6	7	8	9	10
0	1	0	0	0	0	0	0	0	0	0	0
1	1	1	0	0	0	0	0	0	0	0	0
2	$1+\rho$	1	1	0	0	0	0	0	0	0	0
3	$1+2\rho$	$1+2\rho$	1	1	0	0	0	0	0	0	0
4	$1+3\rho+2\rho^2$	$1+3\rho$	$1+3\rho$	1	1	0	0	0	0	0	0
5	$1+4\rho+5\rho^2$	$1+4\rho+5\rho^2$	$1+4\rho$	$1+4\rho$	1	1	0	0	0	0	0
6	$1+5\rho+9\rho^2$ $+5\rho^3$	$1+5\rho+9\rho^2$	$1+5\rho+9\rho^2$	$1+5\rho$	$1+5\rho$	1	1	0	0	0	0
7	$1+6\rho+14\rho^2$ $+14\rho^3$	$1+6\rho+14\rho^2$ $+14\rho^3$	$1+6\rho+14\rho^2$	$1+6\rho+14\rho^2$	$1+6\rho$	$1+6\rho$	1	1	0	0	0
8	$1+7\rho+20\rho^2$ $+28\rho^3+14\rho^4$	$1+7\rho+20\rho^2$ $+28\rho^3$	$1+7\rho+20\rho^2$ $+28\rho^3$	$1+7\rho+20\rho^2$	$1+7\rho+20\rho^2$	$1+7\rho$	$1+7\rho$	1	1	0	0
9	$1+8\rho+27\rho^2$ $+48\rho^3+42\rho^4$	$1+8\rho+27\rho^2$ $+48\rho^3+42\rho^4$	$1+8\rho+27\rho^2$ $+48\rho^3$	$1+8\rho+27\rho^2$ $+48\rho^3$	$1+8\rho+27\rho^2$	$1+8\rho+27\rho^2$	$1+8\rho$	$1+8\rho$	1	1	0
10	$1+9\rho+35\rho^2$ $+75\rho^3+90\rho^4$ $+42\rho^5$	$1+9\rho+35\rho^2$ $+75\rho^3+90\rho^4$	$1+9\rho+35\rho^2$ $+75\rho^3+90\rho^4$	$1+9\rho+35\rho^2$ $+75\rho^3$	$1+9\rho+35\rho^2$ $+75\rho^3$	$1+9\rho+35\rho^2$	$1+9\rho+35\rho^2$	$1+9\rho$	$1+9\rho$	1	1

if m is odd, $m = 2k + 1$,

$$\left[\frac{m}{2}\right] = k, \qquad \left[\frac{m-1}{2}\right] = k, \qquad \left[\frac{m+1}{2}\right] = k + 1$$

Thus in either case we have

$$
\begin{aligned}
a(m,0) + \rho a(m,1) &= \sum_{r=0}^{[(m+1)/2]} \left\{ \binom{m}{r} - \binom{m}{r-1} + \binom{m}{r-1} - \binom{m}{r-2} \right\} \rho^r \\
&= \sum_{r=0}^{[(m+1)/2]} \left\{ \binom{m+1}{r} - \binom{m+1}{r-1} \right\} \rho^r \\
&= a(m+1,0)
\end{aligned}
$$

since

$$\binom{m}{r} + \binom{m}{r-1} = \binom{m+1}{r}$$

by a well-known property of the binomial coefficients.

Thus we have established by induction on m the result of (A10) when $n = 0$. In the same way we can establish the result for all m and non-zero n, so that we may write

$$a(m,n) = \sum_{r=0}^{[(m-n)/2]} \left\{ \binom{m}{r} - \binom{m}{r-1} \right\} \rho^r$$

Thus

$$p_n(\tau) = e^{-(1+\rho)\tau} \rho^n \sum_{m=0}^{\infty} \left[\sum_{r=0}^{[(m-n)/2]} \left\{ \binom{m}{r} - \binom{m}{r-1} \right\} \rho^r \right] \frac{\tau^m}{m!} \qquad \text{(A11)}$$

Again we can rewrite (A11) in the form

$$p_n(\tau) = e^{-(1+\rho)\tau} \rho^n \sum_{m=0}^{\infty} [(1-\rho)(1+\rho)^m + (\rho-1)(1+\rho)^m + a(m,n)] \frac{\tau^m}{m!}$$

$$= (1-\rho)\rho^n + e^{-(1+\rho)\tau} \rho^n \sum_{m=0}^{\infty} [(\rho-1)(1+\rho)^m + a(m,n)] \frac{\tau^m}{m!} \qquad \text{(A12)}$$

Now since $0 \le p_n(\tau) \le 1$ for all τ, then for $\rho < 1$ as $\tau \to \infty$

$$p_n(\tau) \to (1-\rho)\rho^n$$

the well-known steady-state solution (2.6), and (A12) shows this form quite clearly.

The special case $\rho = 1$

When $\rho = 1$

$$a(m,n) = \sum_{r=0}^{[(m-n)/2]} \left\{ \binom{m}{r} - \binom{m}{r-1} \right\} \rho^r$$

becomes

$$\binom{m}{0} - \binom{m}{-1} + \binom{m}{1} - \binom{m}{0} + \binom{m}{2} - \binom{m}{1} + \cdots + \left(\left[\frac{m}{m-n} \right] \right) - \left(\left[\frac{m}{m-n} \right] - 1 \right) = \left(\left[\frac{m}{m-n} \right] \right)$$

Thus

$$p_n(\tau) = e^{-2\tau} \sum_{m=0}^{\infty} \left(\left[\frac{m}{m-n} \atop 2 \right] \right) \frac{\tau^m}{m!}$$

$$= e^{-2\tau} \sum_{m=0}^{\infty} \frac{\tau^m}{\left[\frac{m-n}{2} \right]! \left[\frac{m+n+1}{2} \right]!}$$

$$= e^{-2\tau} \sum_{m=n}^{\infty} \frac{\tau^m}{\left[\frac{m-n}{2} \right]! \left[\frac{m+n+1}{2} \right]!} \tag{A.13}$$

It is left as an exercise to show that

$$m - \left[\frac{m-n}{2} \right] = \left[\frac{m+n+1}{2} \right]$$

Numerical calculations with this formula yielded the same results as before when $\rho = 1$.

$\Sigma_{n=0}^{\infty} p_n(\tau) = 1$ for all τ

Since for each value of τ the number of customers has to be some value of n in $0 \le n \le \infty$,

$$\sum_{n=0}^{\infty} p_n(\tau) = 1, \qquad \text{for all } \tau$$

Of course $p_n(\tau)$ is given in general by (A11) and, because $[(m-n)/2]$ only exists for $m \ge n$, we see that

$$p_n(\tau) = e^{-(1+\rho)\tau} \rho^n \sum_{m=n}^{\infty} \left[\sum_{r=0}^{[(m-n)/2]} \left\{ \binom{m}{r} - \binom{m}{r-1} \right\} \rho^r \right] \frac{\tau^m}{m!}$$

Thus

$$\sum_{n=0}^{\infty} p_n(\tau) = e^{-(1+\rho)\tau} \sum_{n=0}^{\infty} \rho^n \left\{ \sum_{m=n}^{\infty} \left[\sum_{r=0}^{[(m-n)/2]} \left\{ \binom{m}{r} - \binom{m}{r-1} \right\} \rho^r \right] \frac{\tau^m}{m!} \right\}$$

$$= e^{-(1+\rho)\tau} \sum_{m=0}^{\infty} \left[\sum_{n=0}^{m} \sum_{r=0}^{[(m-n)/2]} \left\{ \binom{m}{r} - \binom{m}{r-1} \right\} \rho^{r+n} \right] \frac{\tau^m}{m!}$$

on changing the order of summation for n and m.

Thus, disregarding for the moment the $\exp\{-(1+\rho)\tau\}$, the coefficient of $\tau^m/m!$ in the summation over m is

$$\sum_{n=0}^{m} \sum_{r=0}^{[(m-n)/2]} \left\{ \binom{m}{r} - \binom{m}{r-1} \right\} \rho^{r+n}$$

with region of summation as shown in Fig. A1. If we change the summation variables to $l = n$ and $k = n + r$, then the above becomes

$$\sum_{l=0}^{m} \sum_{k=l}^{[(m+l)/2]} \left\{ \binom{m}{k-l} - \binom{m}{k-l-1} \right\} \rho^k$$

with region of summation as shown in Fig. A2.

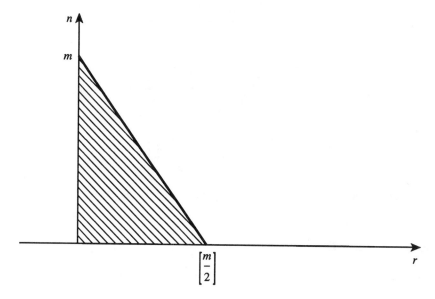

Fig. A1

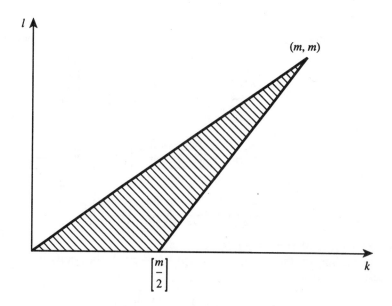

Fig. A2

If we change the order of summation, then from the diagram we see that this expression is

$$\sum_{k=0}^{[m/2]}\sum_{l=0}^{k}\left\{\binom{m}{k-l}-\binom{m}{k-l-1}\right\}\rho^k + \sum_{k=[m/2]+1}^{m}\sum_{l=2k-m}^{k}\left\{\binom{m}{k-l}-\binom{m}{k-l-1}\right\}\rho^k$$

$$=\sum_{k=0}^{[m/2]}\binom{m}{k}\rho^k + \sum_{k=[m/2]+1}^{m}\binom{m}{m-k}\rho^k$$

$$=\sum_{k=0}^{m}\binom{m}{k}\rho^k \quad \text{since} \binom{m}{k}=\binom{m}{m-k}$$

$$=(1+\rho)^m$$

Thus

$$\sum_{n=0}^{\infty}p_n(t)=e^{-(1+\rho)t}\sum_{m=0}^{\infty}\frac{(1+\rho)^m t^m}{m!}=1$$

The preceding manipulations with the changes in order of summation etc. may appear rather obscure. Essentially the coefficient of $t^m/m!$ in $\sum_{n=0}^{\infty}p_n(t)$ is given by $\sum_{n=0}^{m}a(m,n)\rho^n$. Thus from Table A1 of polynomials $a(m,n)$ for $m=5$, for example, by summing across columns 0 to 5, having multiplied the entries by ρ^n, we obtain

$$\rho^0(1+4\rho+5\rho^2)+\rho(1+4\rho+5\rho^2)+\rho^2(1+4\rho)+\rho^3(1+4\rho)+\rho^4.1+\rho^5.1$$
$$=1+5\rho+10\rho^2+10\rho^3+5\rho^4+\rho^5=(1+\rho)^5$$

as required.

Some numerical results

Dr. A.S. Wood of the University of Bradford wrote some FORTRAN programs to evaluate $p_n(\tau)$ using (A11). Some of the results for values of τ $(0\leqslant\tau\leqslant 10)$ and n $(0\leqslant n\leqslant 10)$ are presented in Tables A2–A6. When $\rho<1$ the convergence to the steady state as τ increases is clearly shown. For $\rho\geqslant 1$ the nature of the solution changes. For small τ, the $p_n(\tau)$ are moderate for small values of n only. As τ increases then $p_n(\tau)$ increases with increasing n only eventually to revert to zero as τ increases further. Of course in theory for all finite n, when $\rho\geqslant 1$, $p_n(\tau)\rightarrow 0$, as $\tau\rightarrow\infty$ so that the expected number in the system increases with increasing τ. All the probability moves off to values of n at infinity.

Acknowledgements

During the autumn of 1994 Professor O.P. Sharma from Delhi worked with me at the University of Bradford. His visit was supported by a grant from the Royal Society and from The Indian National Science Academy. We are both of us deeply grateful for that support. The collaborative work carried out on that visit produced the simple solution given above and means that it is now possible to present this work at the student level. Professor Sharma deserves full credit for the joint work presented here.

References

Bailey, N.T.J. (1954) A continuous time treatment of a simple queue using generating functions. *Journal of the Royal Statistics Society* **B16**, 288–91.

Brockmeyer, E., Halstrøm, H.L. and Jensen, A. (1948) The life and works of A.K. Erlang. *Transactions of the Danish Academy of Technology and Science* **2**.

Champernowne, D.G. (1956) An elementary method of solution of the queueing problem with a single server and a constant parameter. *Journal of the Royal Statistics Society* **B18**, 125–8.

Clarke, A.B. (1953) *The time dependent waiting line problem.* The University of Michigan Report M720-1R39.

Conolly, B.W. and Langaris, C. (1993) On a new formula for the transient state probabilities for M/M/1 queues and computational implications. *Journal of Applied Probability* **30**, 237–46.

Kleinrock, L. (1975) *Queueing systems*, Vol. 1. New York: John Wiley and Sons.

Sharma, O.P. (1990) *Markovian queues.* Chichester: Ellis Horwood.

Examples A1

1　Use the induction method in the text to prove that for $n \neq 0$ the assumption of (A10) for values up to m leads to the correct form for $a(m+1, n)$ on using (A8).

2　Verify the correctness of formula (A13).

3　Use (A13) to show that in the special case $\rho = 1$,

$$\sum_{n=0}^{\infty} p_n(\tau) = 1 \qquad \text{for all } \tau$$

4　Use the method of induction to show that

$$\sum_{n=0}^{\infty} a(m, n)\rho^n = \sum_{n=0}^{m} a(m, n)\rho^n = (1 + \rho)^m$$

As discussed in the text, this result is needed to establish that

$$\sum_{n=0}^{\infty} p_n(\tau) = 1 \qquad \text{for all } \tau$$

Solutions to Examples A1

1　On the R.H.S. of (A8) we have

$$a(m + 1, n - 1) + \rho a(m + 1, n + 1)$$

$$= \sum_{r=0}^{[(m-n+1)/2]} \left\{ \binom{m}{r} - \binom{m}{r-1} \right\} \rho^r + \sum_{r=0}^{[(m-n-1)/2]} \left\{ \binom{m}{r} - \binom{m}{r-1} \right\} \rho^{r+1}$$

If $m - n = 2k$ (even),

$$\left[\frac{m-n+1}{2} \right] = k, \qquad \left[\frac{m-n-1}{2} \right] = k - 1$$

If $m - n = 2k + 1$ (odd),

$$\left[\frac{m - n + 1}{2}\right] = k + 1, \qquad \left[\frac{m - n - 1}{2}\right] = k$$

Thus in either case

$$a(m + 1, n - 1) + \rho a(m + 1, n + 1)$$

$$= \sum_{r=0}^{[(m+1-n)/2]} \left\{ \binom{m}{r} + \binom{m}{r - 1} - \binom{m}{r - 1} - \binom{m}{r - 2} \right\} \rho^r$$

$$= \sum_{r=0}^{[(m+1-n)/2]} \left\{ \binom{m + 1}{r} - \binom{m + 1}{r - 1} \right\} \rho^r$$

$$= a(m + 1, n)$$

as required.

2 We have

$$p_n(\tau) = e^{-2t} \sum_{m=0}^{\infty} \left(\left[\frac{m}{\dfrac{m - n}{2}}\right] \right) \frac{\tau^m}{m!}$$

$$= e^{-2\tau} \sum_{m=0}^{\infty} \frac{\tau^m}{\left[\dfrac{m - n}{2}\right]! \left(m - \left[\dfrac{m - n}{2}\right]\right)!}$$

There are four cases to consider: m and n even or odd.
If

$$m = 2k + 1, \, n = 2l + 1, \, \left[\frac{m - n}{2}\right] = k - l \quad \text{and} \quad m - \left[\frac{m - n}{2}\right] = k + l + 1 = \left[\frac{m + n + 1}{2}\right]$$

$$m = 2k + 1, \, n = 2l, \, \left[\frac{m - n}{2}\right] = k - l \quad \text{and} \quad m - \left[\frac{m - n}{2}\right] = k + l + 1 = \left[\frac{m + n + 1}{2}\right]$$

$$m = 2k, \, n = 2l + 1, \, \left[\frac{m - n}{2}\right] = k - l - 1 \quad \text{and} \quad m - \left[\frac{m - n}{2}\right] = k + l + 1 = \left[\frac{m + n + 1}{2}\right]$$

$$m = 2k, \, n = 2l, \, \left[\frac{m - n}{2}\right] = k - l \quad \text{and} \quad m - \left[\frac{m - n}{2}\right] = k + l = \left[\frac{m + n + 1}{2}\right]$$

In each case

$$m - \left[\frac{m - n}{2}\right] = \left[\frac{m + n + 1}{2}\right]$$

so

$$p_n(\tau) = e^{-2\tau} \sum_{m=0}^{\infty} \frac{\tau^m}{\left[\dfrac{m-n}{2}\right]! \left[\dfrac{m+n+1}{2}\right]!}$$

3 When $\rho = 1$,

$$p_n(\tau) = e^{-2\tau} \sum_{m=n}^{\infty} \left(\left[\dfrac{m}{\dfrac{m-n}{2}} \right] \right) \frac{\tau^m}{m!}$$

$$\therefore \sum_{n=0}^{\infty} p_n(\tau) = e^{-2\tau} \sum_{n=0}^{\infty} \sum_{m=n}^{\infty} \left(\left[\dfrac{m}{\dfrac{m-n}{2}} \right] \right) \frac{\tau^m}{m!}$$

where the summation is over the region shown in Fig. A3

$$= e^{-2\tau} \sum_{m=0}^{\infty} \left[\sum_{n=0}^{m} \left(\left[\dfrac{m}{\dfrac{m-n}{2}} \right] \right) \right] \frac{\tau^m}{m!}$$

on changing the order of summation. But

$$\sum_{n=0}^{m} \left(\left[\dfrac{m}{\dfrac{m-n}{2}} \right] \right) = \sum_{n=0}^{m} \binom{m}{n} = 2^m$$

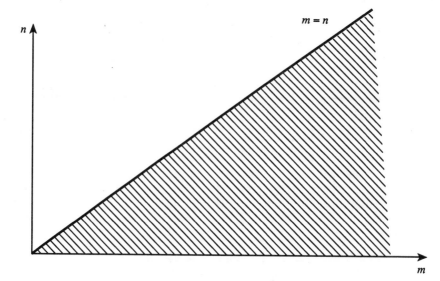

Fig. A3

There are two cases to consider, m even or m odd. If $m = 2k$ (even), then

$$\sum_{n=0}^{m} \binom{m}{\left[\dfrac{m-n}{2}\right]} = \underset{n=0}{\binom{2k}{k}} + \underset{n=1}{\binom{2k}{k-1}} + \underset{n=2}{\binom{2k}{k-1}} + \underset{n=3}{\binom{2k}{k-2}} + \cdots + \underset{n=2k-1}{\binom{2k}{0}} + \underset{n=2k}{\binom{2k}{0}}$$

$$= \binom{2k}{0} + \binom{2k}{1} + \binom{2k}{2} + \cdots + \binom{2k}{k-1} + \binom{2k}{k} + \binom{2k}{k+1}$$

$$+ \cdots + \binom{2k}{2k-1} + \binom{2k}{2k}$$

$$= 2^{2k}$$

since

$$\binom{2k}{r} = \binom{2k}{2k-r}$$

Similarly if $m = 2k + 1$ (odd), then

$$\sum_{n=0}^{m} \binom{m}{\left[\dfrac{m-n}{2}\right]} = \underset{n=0}{\binom{2k+1}{k}} + \underset{n=1}{\binom{2k+1}{k}} + \underset{n=2}{\binom{2k+1}{k-1}} + \underset{n=3}{\binom{2k+1}{k-1}}$$

$$+ \cdots + \underset{n=2k}{\binom{2k+1}{0}} + \underset{n=2k+1}{\binom{2k+1}{0}}$$

$$= \binom{2k+1}{0} + \binom{2k+1}{1} + \binom{2k+1}{2}$$

$$+ \cdots + \binom{2k+1}{2k-1} + \binom{2k+1}{2k} + \binom{2k+1}{2k+1}$$

$$= 2^{2k+1}$$

for the same reason.
Thus

$$\sum_{n=0}^{\infty} p_n(\tau) = e^{-2\tau} \sum_{m=0}^{\infty} \frac{2^m \tau^m}{m!} = e^{-2\tau} e^{2\tau} = 1$$

4 Assume that $\sum_{n=0}^{\infty} \rho^n a(m, n) = (1 + \rho)^m$ for some value of m. Of course, since $a(m, n) = 0$ if $n > m$, then the above sum is effectively from $n = 0$ to $n = m$. In the case when $m = 1$, $a(1, 0) = 1$ and $a(1, 1) = 1$, so that

$$\rho^0 a(1, 0) + \rho a(1, 1) = 1 + \rho = (1 + \rho)^1$$

When $m = 2$, $a(2, 0) = 1 + \rho$, $a(2, 1) = 1$, $a(2, 2) = 1$.

$$\therefore \quad \rho^0 a(2, 0) + \rho a(2, 1) + \rho^2 a(2, 2) = 1 + \rho + \rho + \rho^2 = (1 + \rho)^2$$

Thus the result is certainly true for $m = 1$ or $m = 2$.

We assume that $\sum_{n=0}^{m} \rho^n a(m, n) = (1 + \rho)^m$. Then

$$\sum_{n=0}^{m+1} \rho^n a(m+1, n)$$

$$= a(m+1, 0) + \sum_{n=1}^{m} \rho^n a(m+1, n) + \rho^{m+1} a(m+1, m+1)$$

$$= a(m, 0) + \rho a(m, 1) + \sum_{n=1}^{m} \rho^n [a(m, n-1) + \rho a(m, n+1)] + \rho^{m+1}$$

$$= \sum_{n=0}^{m} \rho^n a(m, n) + \rho \sum_{n=0}^{m} \rho^n a(m, n)$$

since $a(m, m+1) = 0$ and $a(m, m) = 1$

$$= (1 + \rho)(1 + \rho)^m = (1 + \rho)^{m+1}$$

as required. Hence the result is established by induction.

Final update

The formula (A.11) for $p_n(\tau)$ as given by Bunday and Sharma (1995) can be generalised and extended to cover the case when there are i in the system at the outset (O'Neill and Bunday, 1995). In that case

$$p_n^{(i)}(\tau) = e^{-(1+\rho)\tau} \rho^{n-i} \left[\sum_{m=|n-i|}^{n+i} \binom{m}{l} \rho^l \frac{\tau^m}{m!} + \sum_{m=n+i+1}^{\infty} \left\{ \sum_{r=i}^{[l]} A(m, n, i, r) \rho^r \right\} \frac{\tau^m}{m!} \right] \quad \text{(A.14)}$$

where

$$l = \frac{1}{2}(m - n + i), \quad A(m, n, i, r) = \binom{m}{r-i} - \binom{m}{r-i-1} \text{ if } r < l$$

$$= \binom{m}{r} - \binom{m}{r-i-1} \text{ if } r = l.$$

Note that if m is a positive integer $\binom{m}{k} = 0$ unless k is a positive integer such that $0 \leqslant k \leqslant m$. It is not difficult to show that in the special case $i = 0$, (A.14) becomes (A.11). The polynomial in ρ, which constitutes the coefficient of $\tau^m/m!$, shows the same structure involving the difference of Binomial coefficients as was apparent in (A.11).

References

Bunday, B.D. and Sharma, O.P. (1995). *A simple formula for the transient state probabilities of an M/M/1/∞ queue*. Report No. SOR 95–66, Department of Mathematics, University of Bradford.

O'Neill, P.D. and Bunday, B.D. (1995). *The transient state probabilities of an M/M/1/∞ queue via a random walk approach*. Report No. SOR 95–67, Department of Mathematics, University of Bradford.

Table A2 $p_n(\tau)$ for $\rho = 0.1$

τ	0	1	n 2	3	4	5
0.0	1.0000000	0.0000000	0.0000000	0.0000000	0.0000000	0.0000000
0.5	0.9614018	0.0377210	0.0008632	0.0000138	0.0000002	0.0000000
1.0	0.9385969	0.0588803	0.0024470	0.0000741	0.0000017	0.0000000
1.5	0.9247913	0.0710310	0.0040001	0.0001715	0.0000059	0.0000002
2.0	0.9162373	0.0781746	0.0052907	0.0002844	0.0000124	0.0000005
2.5	0.9108212	0.0824723	0.0062890	0.0003959	0.0000207	0.0000009
3.0	0.9073229	0.0851157	0.0070335	0.0004965	0.0000298	0.0000015
3.5	0.9050224	0.0867759	0.0075781	0.0005823	0.0000389	0.0000023
4.0	0.9034852	0.0878388	0.0079725	0.0006529	0.0000474	0.0000030
4.5	0.9024433	0.0885315	0.0082568	0.0007094	0.0000550	0.0000038
5.0	0.9017282	0.0889901	0.0084613	0.0007539	0.0000616	0.0000046
5.5	0.9012320	0.0892981	0.0086086	0.0007885	0.0000672	0.0000053
6.0	0.9008843	0.0895076	0.0087148	0.0008152	0.0000718	0.0000059
6.5	0.9006386	0.0896517	0.0087915	0.0008356	0.0000756	0.0000064
7.0	0.9004636	0.0897518	0.0088472	0.0008512	0.0000787	0.0000069
7.5	0.9003382	0.0898220	0.0088877	0.0008630	0.0000811	0.0000073
8.0	0.9002478	0.0898716	0.0089172	0.0008720	0.0000831	0.0000076
8.5	0.9001823	0.0899069	0.0089388	0.0008788	0.0000846	0.0000079
9.0	0.9001345	0.0899322	0.0089546	0.0008839	0.0000858	0.0000081
9.5	0.9000996	0.0899504	0.0089663	0.0008878	0.0000867	0.0000083
10.0	0.9000740	0.0899636	0.0089749	0.0008908	0.0000875	0.0000084

τ	6	7	8	9	10
0.0	0.0000000	0.0000000	0.0000000	0.0000000	0.0000000
0.5	0.0000000	0.0000000	0.0000000	0.0000000	0.0000000
1.0	0.0000000	0.0000000	0.0000000	0.0000000	0.0000000
1.5	0.0000000	0.0000000	0.0000000	0.0000000	0.0000000
2.0	0.0000000	0.0000000	0.0000000	0.0000000	0.0000000
2.5	0.0000000	0.0000000	0.0000000	0.0000000	0.0000000
3.0	0.0000001	0.0000000	0.0000000	0.0000000	0.0000000
3.5	0.0000001	0.0000000	0.0000000	0.0000000	0.0000000
4.0	0.0000002	0.0000000	0.0000000	0.0000000	0.0000000
4.5	0.0000002	0.0000000	0.0000000	0.0000000	0.0000000
5.0	0.0000003	0.0000000	0.0000000	0.0000000	0.0000000
5.5	0.0000004	0.0000000	0.0000000	0.0000000	0.0000000
6.0	0.0000004	0.0000000	0.0000000	0.0000000	0.0000000
6.5	0.0000005	0.0000000	0.0000000	0.0000000	0.0000000
7.0	0.0000006	0.0000000	0.0000000	0.0000000	0.0000000
7.5	0.0000006	0.0000000	0.0000000	0.0000000	0.0000000
8.0	0.0000007	0.0000001	0.0000000	0.0000000	0.0000000
8.5	0.0000007	0.0000001	0.0000000	0.0000000	0.0000000
9.0	0.0000007	0.0000001	0.0000000	0.0000000	0.0000000
9.5	0.0000008	0.0000001	0.0000000	0.0000000	0.0000000
10.0	0.0000008	0.0000001	0.0000000	0.0000000	0.0000000

Table A3 $p_n(\tau)$ for $\rho = 0.5$

τ	0	1	2	3	4	5
0.0	1.00000000	0.0000000	0.0000000	0.0000000	0.0000000	0.0000000
0.5	0.8210865	0.1592878	0.0180991	0.0014352	0.0000870	0.0000043
1.0	0.7262551	0.2212476	0.0449433	0.0066932	0.0007800	0.0000744
1.5	0.6701340	0.2466142	0.0667124	0.0138663	0.0023115	0.0003197
2.0	0.6337954	0.2572818	0.0824007	0.0211368	0.0044553	0.0007909
2.5	0.6085790	0.2616628	0.0934172	0.0276657	0.0068873	0.0014656
3.0	0.5901459	0.2632093	0.1011850	0.0332168	0.0093564	0.0022851
3.5	0.5761330	0.2634313	0.1067442	0.0378234	0.0117102	0.0031857
4.0	0.5651567	0.2630282	0.1107952	0.0416115	0.0138730	0.0041141
4.5	0.5563557	0.2623397	0.1138010	0.0447226	0.0158172	0.0050315
5.0	0.5491661	0.2615368	0.1160691	0.0472847	0.0175429	0.0059127
5.5	0.5432032	0.2607070	0.1178069	0.0494051	0.0190638	0.0067427
6.0	0.5381952	0.2598951	0.1191565	0.0511700	0.0203994	0.0075141
6.5	0.5339445	0.2591233	0.1202173	0.0526480	0.0215708	0.0082245
7.0	0.5303039	0.2584016	0.1210600	0.0538934	0.0225983	0.0088746
7.5	0.5271616	0.2577332	0.1217359	0.0549489	0.0235005	0.0094670
8.0	0.5244309	0.2571179	0.1222825	0.0558486	0.0242941	0.0100054
8.5	0.5220439	0.2565534	0.1227278	0.0566194	0.0249935	0.0104938
9.0	0.5199462	0.2560364	0.1230930	0.0572833	0.0256112	0.0109365
9.5	0.5180941	0.2555636	0.1233944	0.0578575	0.0261581	0.0113376
10.0	0.5164520	0.2551313	0.1236444	0.0583565	0.0266433	0.0117010

τ	6	7	8	9	10
0.0	0.0000000	0.0000000	0.0000000	0.0000000	0.0000000
0.5	0.0000002	0.0000000	0.0000000	0.0000000	0.0000000
1.0	0.0000060	0.0000004	0.0000000	0.0000000	0.0000000
1.5	0.0000377	0.0000039	0.0000003	0.0000000	0.0000000
2.0	0.0001208	0.0000161	0.0000019	0.0000002	0.0000000
2.5	0.0002709	0.0000441	0.0000064	0.0000008	0.0000001
3.0	0.0004897	0.0000931	0.0000159	0.0000024	0.0000003
3.5	0.0007678	0.0001654	0.0000321	0.0000057	0.0000009
4.0	0.0010910	0.0002604	0.0000563	0.0000111	0.0000020
4.5	0.0014438	0.0003755	0.0000890	0.0000193	0.0000039
5.0	0.0018123	0.0005067	0.0001298	0.0000306	0.0000067
5.5	0.0021851	0.0006499	0.0001779	0.0000450	0.0000106
6.0	0.0025535	0.0008010	0.0002324	0.0000625	0.0000157
6.5	0.0029112	0.0009564	0.0002919	0.0000830	0.0000220
7.0	0.0032542	0.0011130	0.0003552	0.0001059	0.0000296
7.5	0.0035797	0.0012685	0.0004211	0.0001311	0.0000383
8.0	0.0038863	0.0014208	0.0004885	0.0001580	0.0000481
8.5	0.0041736	0.0015686	0.0005565	0.0001863	0.0000589
9.0	0.0044415	0.0017109	0.0006242	0.0002156	0.0000705
9.5	0.0046906	0.0018470	0.0006910	0.0002454	0.0000828
10.0	0.0049217	0.0019766	0.0007565	0.0002756	0.0000955

Table A4 $p_n(\tau)$ for $\rho = 1$

τ	0	1	2	3	4	5
0.0	1.0000000	0.0000000	0.0000000	0.0000000	0.0000000	0.0000000
0.5	0.6736700	0.2578492	0.0580941	0.0091622	0.0011068	0.0001081
1.0	0.5237776	0.3085083	0.1220303	0.0356566	0.0081951	0.0015463
1.5	0.4398271	0.3086093	0.1595659	0.0639992	0.0207568	0.0056205
2.0	0.3857528	0.2963773	0.1787508	0.0870643	0.0351843	0.0120735
2.5	0.3475131	0.2819242	0.1875626	0.1040298	0.0489593	0.0198787
3.0	0.3187089	0.2680251	0.1907093	0.1159736	0.0609898	0.0280700
3.5	0.2960270	0.2553729	0.1907537	0.1241794	0.0710258	0.0360024
4.0	0.2775743	0.2440387	0.1890906	0.1296948	0.0791944	0.0433268
4.5	0.2621845	0.2339123	0.1864957	0.1332901	0.0857508	0.0498962
5.0	0.2490960	0.2248435	0.1834112	0.1355129	0.0909669	0.0556826
5.5	0.2377879	0.2166865	0.1800942	0.1367494	0.0950884	0.0607207
6.0	0.2278905	0.2093131	0.1766969	0.1372728	0.0983230	0.0650745
6.5	0.2191332	0.2026144	0.1733092	0.1372782	0.1008406	0.0688178
7.0	0.2113129	0.1964990	0.1699843	0.1369059	0.1027785	0.0720245
7.5	0.2042737	0.1908905	0.1667528	0.1362582	0.1042466	0.0747633
8.0	0.1978937	0.1857250	0.1636312	0.1354103	0.1053330	0.0770965
8.5	0.1920762	0.1809489	0.1606273	0.1344178	0.1061084	0.0790784
9.0	0.1867431	0.1765168	0.1577435	0.1333223	0.1066295	0.0807569
9.5	0.1818307	0.1723901	0.1549787	0.1321551	0.1069423	0.0821729
10.0	0.1772865	0.1685359	0.1523300	0.1309399	0.1070840	0.0833618

τ	6	7	8	9	10
0.0	0.0000000	0.0000000	0.0000000	0.0000000	0.0000000
0.5	0.0000089	0.0000006	0.0000000	0.0000000	0.0000000
1.0	0.0002470	0.0000342	0.0000042	0.0000005	0.0000000
1.5	0.0013022	0.0002632	0.0000471	0.0000076	0.0000011
2.0	0.0035861	0.0009367	0.0002180	0.0000457	0.0000087
2.5	0.0070666	0.0022276	0.0006295	0.0001610	0.0000376
3.0	0.0114337	0.0041640	0.0013682	0.0004089	0.0001119
3.5	0.0163121	0.0066599	0.0024685	0.0008363	0.0002605
4.0	0.0213777	0.0095741	0.0039158	0.0014709	0.0005101
4.5	0.0263934	0.0127570	0.0056624	0.0023191	0.0008803
5.0	0.0312046	0.0160758	0.0076447	0.0033692	0.0013814
5.5	0.0357211	0.0194241	0.0097961	0.0045975	0.0020146
6.0	0.0398981	0.0227228	0.0120548	0.0059741	0.0027735
6.5	0.0437210	0.0259165	0.0143674	0.0074668	0.0036467
7.0	0.0471941	0.0289688	0.0166904	0.0090445	0.0046193
7.5	0.0503326	0.0318580	0.0189899	0.0106787	0.0056751
8.0	0.0531580	0.0345730	0.0212400	0.0123445	0.0067978
8.5	0.0556943	0.0371100	0.0234221	0.0140210	0.0079714
9.0	0.0579661	0.0394706	0.0255233	0.0156908	0.0091815
9.5	0.0599975	0.0416601	0.0275349	0.0173400	0.0104151
10.0	0.0618114	0.0436856	0.0294523	0.0189576	0.0116608

Table A5 $p_n(\tau)$ for $\rho = 1.5$

τ	0	1	2	3	4	5
0.0	1.0000000	0.0000000	0.0000000	0.0000000	0.0000000	0.0000000
0.5	0.5523181	0.3129439	0.1048716	0.0246728	0.0044536	0.0006509
1.0	0.3754169	0.3214523	0.1859824	0.0800336	0.0272217	0.0076262
1.5	0.2836922	0.2860461	0.2129405	0.1239532	0.0587644	0.0233871
2.0	0.2272509	0.2493646	0.2139758	0.1492679	0.0870800	0.0434353
2.5	0.1886845	0.2179402	0.2045955	0.1607814	0.1080067	0.0630613
3.0	0.1604948	0.1918731	0.1915681	0.1636830	0.1216921	0.0797525
3.5	0.1389142	0.1702385	0.1777513	0.1615489	0.1295933	0.0927377
4.0	0.1218299	0.1521235	0.1643337	0.1565931	0.1332425	0.1021540
4.5	0.1079571	0.1367947	0.1517804	0.1501446	0.1338937	0.1084959
5.0	0.0964657	0.1236900	0.1402368	0.1429954	0.1324861	0.1123366
5.5	0.0867937	0.1123809	0.1297066	0.1356156	0.1296973	0.1142091
6.0	0.0785452	0.1025390	0.1201331	0.1282826	0.1260088	0.1145667
6.5	0.0714332	0.0939096	0.1114370	0.1211567	0.1217597	0.1137777
7.0	0.0652435	0.0862927	0.1035338	0.1143269	0.1171881	0.1121348
7.5	0.0598132	0.0795296	0.0963422	0.1078380	0.1124605	0.1098670
8.0	0.0550161	0.0734926	0.0897870	0.1017083	0.1076928	0.1071522
8.5	0.0507522	0.0680780	0.0838006	0.0959391	0.1029650	0.1041276
9.0	0.0469420	0.0632008	0.0783228	0.0905223	0.0983314	0.1008991
9.5	0.0435207	0.0587904	0.0733004	0.0854437	0.0938285	0.0975476
10.0	0.0404354	0.0547882	0.0686865	0.0806861	0.0894793	0.0941349

τ	6	7	8	9	10
0.0	0.0000000	0.0000000	0.0000000	0.0000000	0.0000000
0.5	0.0000798	0.0000084	0.0000008	0.0000001	0.0000000
1.0	0.0018125	0.0003736	0.0000679	0.0000110	0.0000016
1.5	0.0079967	0.0023922	0.0006353	0.0001515	0.0000328
2.0	0.0188579	0.0072314	0.0024790	0.0007675	0.0002165
2.5	0.0324466	0.0148847	0.0061490	0.0023072	0.0007922
3.0	0.0465773	0.0244695	0.0116587	0.0050743	0.0020303
3.5	0.0597247	0.0348813	0.0185998	0.0091101	0.0041208
4.0	0.0710897	0.0451914	0.0263908	0.0142302	0.0071176
4.5	0.0803930	0.0547670	0.0344664	0.0201246	0.0109456
5.0	0.0876602	0.0632487	0.0423702	0.0264515	0.0154422
5.5	0.0930703	0.0704805	0.0497792	0.0328973	0.0204035
6.0	0.0968626	0.0764414	0.0564907	0.0392060	0.0256212
6.5	0.0992868	0.0811908	0.0623982	0.0451865	0.0309055
7.0	0.1005772	0.0848314	0.0674648	0.0507082	0.0360976
7.5	0.1009423	0.0874846	0.0717011	0.0556915	0.0410731
8.0	0.1005618	0.0892756	0.0751477	0.0600965	0.0457404
8.5	0.0995878	0.0903249	0.0778630	0.0639130	0.0500375
9.0	0.0981470	0.0907435	0.0799137	0.0671523	0.0539263
9.5	0.0963441	0.0906316	0.0813695	0.0698395	0.0573886
10.0	0.0942655	0.0900777	0.0822992	0.0720090	0.0604209

Table A6 $p_n(\tau)$ for $\rho = 2$

τ	0	1	2	3	4	5
0.0	1.0000000	0.0000000	0.0000000	0.0000000	0.0000000	0.0000000
0.5	0.4525102	0.3375006	0.1495574	0.0466594	0.0111870	0.0021738
1.0	0.2675907	0.2967181	0.2235147	0.1260131	0.0564053	0.0208605
1.5	0.1802917	0.2331291	0.2229010	0.1677904	0.1035159	0.0538760
2.0	0.1303134	0.1824261	0.1991012	0.1776254	0.1334009	0.0861698
2.5	0.0983322	0.1444794	0.1713645	0.1707318	0.1462817	0.1096029
3.0	0.0763904	0.1159707	0.1456128	0.1566934	0.1474488	0.1230202
3.5	0.0606078	0.0942141	0.1233022	0.1404181	0.1416874	0.1282047
4.0	0.0488619	0.0773335	0.1044549	0.1242273	0.1322891	0.1275096
4.5	0.0398924	0.0640382	0.0886748	0.1091375	0.1213020	0.1229780
5.0	0.0329041	0.0534292	0.0754886	0.0955260	0.1099336	0.1161479
5.5	0.0273707	0.0448681	0.0644568	0.0834628	0.0988633	0.1080991
6.0	0.0229309	0.0378919	0.0552029	0.0728771	0.0884494	0.0995569
6.5	0.0193290	0.0321592	0.0474151	0.0636388	0.0788590	0.0909915
7.0	0.0163794	0.0274134	0.0408383	0.0556001	0.0701468	0.0826955
7.5	0.0139443	0.0234591	0.0352649	0.0486147	0.0623040	0.0748412
8.0	0.0119199	0.0201453	0.0305259	0.0425468	0.0552870	0.0675202
8.5	0.0102266	0.0173541	0.0264834	0.0372747	0.0490349	0.0607717
9.0	0.0088026	0.0149921	0.0230246	0.0326910	0.0434802	0.0546007
9.5	0.0075992	0.0129853	0.0200569	0.0287022	0.0385544	0.0489915
10.0	0.0065780	0.0112738	0.0175037	0.0252275	0.0341916	0.0439154

τ	6	7	8	9	10
0.0	0.0000000	0.0000000	0.0000000	0.0000000	0.0000000
0.5	0.0003548	0.0000499	0.0000062	0.0000007	0.0000001
1.0	0.0065591	0.0017913	0.0004320	0.0000932	0.0000182
1.5	0.0241812	0.0095225	0.0033361	0.0010516	0.0003011
2.0	0.0486856	0.0243909	0.0109587	0.0044579	0.0016553
2.5	0.0727657	0.0432738	0.0232656	0.0113981	0.0051235
3.0	0.0919809	0.0621776	0.0382878	0.0216187	0.0112580
3.5	0.1049690	0.0783423	0.0536336	0.0338667	0.0198214
4.0	0.1121261	0.0905230	0.0674559	0.0466144	0.0299969
4.5	0.1145384	0.0985513	0.0787010	0.0585680	0.0407632
5.0	0.1134039	0.1028425	0.0869845	0.0688620	0.0511866
5.5	0.1097863	0.1040489	0.0923675	0.0770506	0.0605654
6.0	0.1045416	0.1028599	0.0951583	0.0830134	0.0684615
6.5	0.0983209	0.0999053	0.0957718	0.0868468	0.0746684
7.0	0.0916008	0.0957173	0.0946428	0.0887722	0.0791557
7.5	0.0847198	0.0907241	0.0921783	0.0890692	0.0820131
8.0	0.0779107	0.0852593	0.0887351	0.0880310	0.0834022
8.5	0.0713281	0.0795748	0.0846130	0.0859379	0.0835204
9.0	0.0650694	0.0738564	0.0800560	0.0830427	0.0825759
9.5	0.0591912	0.0682372	0.0752573	0.0795649	0.0807710
10.0	0.0537213	0.0628095	0.0703668	0.0756895	0.0782927

Index